Climate Change, Forests and Federalism

Evgeny Guglyuvatyy

Climate Change, Forests and Federalism

Australian Experience

 Springer

Evgeny Guglyuvatyy
Southern Cross University
Bilinga, QLD, Australia

ISBN 978-981-19-0741-8   ISBN 978-981-19-0742-5   (eBook)
https://doi.org/10.1007/978-981-19-0742-5

This Springer imprint is published by the registered company Springer Nature Singapore Pte Ltd.
The registered company address is: 152 Beach Road, #21-01/04 Gateway East, Singapore 189721,
Singapore

Preface

This book examines major existing and emerging climate change laws and policies. A discussion of the international climate regime, as well as key policy approaches to tackling climate change, is provided as a background for reviewing Australian climate regime. The evolution of the Australian approach to climate change and the development of forest law are examined from a historical perspective. The historical approach enables the identification and tracking of policy issues in Australia as a federal state. Through an analysis of climate policy and forest legislation at the federal and state levels, this book provides a legal and empirical understanding of the inconsistencies and challenges of the current regime and considers specific measures to strengthen the climate policy framework in Australia.

Bilinga, QLD, Australia

Evgeny Guglyuvatyy

Contents

Chapter 1
Climate Change and International Response

Abstract The phenomenon of climate change is currently one of the most serious global problems confronting humankind. The potential environmental and economic impacts of climate change are enormous and finding a solution is a complicated task. Notwithstanding the encouraging growth in the number of international environmental treaties over the last 25 years, there are few climate change related treaties at the international level. This chapter discusses the international climate change regime that comprises the United Nations Framework Convention on Climate Change (UNFCCC), the Kyoto Protocol and the Paris agreement to lay the ground for Australian climate policy analysis.

Keywords Climate change and international response · Kyoto Protocol · Paris agreement

1.1 Introduction

Climate change is one of the most serious global challenges facing humankind. Climate change has enormous environmental and economic implications, and finding a solution is a daunting task. Australia, similar to many other countries, is experiencing substantial climate change as a result of global greenhouse gas (GHG) emissions.[1]

Appropriate responses to climate change require coordinated and decisive efforts by governments around the world, as well as major changes in business and public behaviour. From a production side, it requires a shift from fossil fuel-based economy to carbon neutral. Many countries including Australia are dependent on energy-intensive transport and the use of fossil fuels for electricity generation. Against this

---

[1] The primary greenhouse gases in Earth's atmosphere are water vapor (H_2O), carbon dioxide (CO_2), methane (CH_4), nitrous oxide (N_2O), ozone (O_3), Chlorofluorocarbons (CFCs) and Hydrofluorocarbons (includes HCFCs and HFCs).

background, there are many opportunities for the world to decrease its GHG emissions.

Deforestation and forest degradation are also direct contributors to climate change, accounting for about 11% of greenhouse gas emissions.[2] Deforestation occurs at an alarming rate in Australia and internationally, driven by agriculture, mining, infrastructure and other human activities.[3] Emissions from forestry need to be significantly reduced in addition to other mitigation actions to achieve significant reductions in global greenhouse gas emissions and stop climate change.

Australia is presently facing one of the most serious environmental threats in the form of human conducted deforestation. This activity causes significant detrimental impacts upon the environment. The Australian efforts to reduce greenhouse gas emissions included substantial reliance on land use, land-use change and forestry (LULUCF). Australia's ability to meet its emission reduction targets has depended on LULUCF for many years. Hence, development and current state of Australia's climate change policy needs to be examined with reference to the forest policy to assess the most significant aspects and effectiveness of the current regime.

The purpose of this book is to look at the global problem of climate change through the prism of an individual country's attempt to tackle this problem. This book begins with a discussion of the origins of climate change and the evolution of the international response to climate change. Key climate change mitigation actions and policies are considered to provide the necessary framework for analysing Australia's approach to climate change. Australia's climate change policy development is considered from a historical perspective. The final chapter of the book focuses on forests management and protection, and the disharmony between Federal and State governments' approaches to climate change and forests policies.

The book traces the evolution of the response to climate change, focusing on Australia as one of the Federal countries unable to adequately reduce greenhouse gas emissions due to the systematic failure of the Australian government to develop a common and effective approach to the problem of climate change.

1.2   Climate Change Origin

Climate change is the result of an increase in global temperature (global warming) caused by the release of greenhouse gases into the atmosphere.[4] Concern about the intensification of the greenhouse gas effect grew in the 1970s when it was detected that the concentration of greenhouse gases in the atmosphere was steadily rising.

---

[2] European Commission. Forests and Agriculture. https://ec.europa.eu/clima/eu-action/forests-and-agriculture_en

[3] World Wide Fund for Nature and International Institute for Applied Systems Analysis, Great Barrier Reef 2018 WWF – World Wildlife Fund. Available at: http://www.wwf.org.au/what-we-do/oceans/great-barrier-reef#gs.NI0qrXc

[4] International Panel on Climate Change (IPCC) 2021. Sixth Assessment Report.

The Earth receives the energy of the Sun in the form of light. Since the atmosphere is transparent to light, most of the solar energy reaches the Earth and is then absorbed by the surface. Because of its temperature, the Earth's surface emits energy in the infrared range. GHGs are not transparent to infrared radiation, consequently infrared radiation is absorbed and transferred as heat to all gases in the atmosphere. The atmosphere also emits infrared radiation due to its temperature, just like the surface of the Earth. Thus, the surface and lower atmosphere is heated due to GHG emissions.[5]

An increase in the concentration of greenhouse gas emissions leads to an increase in the average temperature of the Earth's surface.[6] Natural sources of carbon dioxide, such as water vapor, are estimated to be 20 times larger than those associated with human activities, but natural sources are balanced by natural sinks, such as photosynthesis of carbon compounds by plants and ocean plankton. A wide range of human activities, including industry, transport, residential heating/cooling, commerce and agriculture, have increased the concentration of CO_2 and several other greenhouse gases.[7] Especially, some of the greenhouse gases, such as hydrofluorocarbons, perfluorocarbons and sulphur hexafluoride, originate exclusively from human production processes. According to the Intergovernmental Panel on Climate Change (IPCC), about half of carbon dioxide (CO_2) emissions between 1750 and 2011 occurred in the last 40 years. CO_2 emissions from fossil fuel combustion added about 78% of the total upsurge in greenhouse gas emissions from 1970 to 2010.[8]

Despite the general consensus of scientists and environmentalists that the increase in greenhouse gas emissions is the result of human activities, and the increase in temperature will have a negative impact on the world, there is still a small minority of climate change sceptics.[9] The sceptics argue that due to deficient data collection systems in the past, there is no definitive way to prove that an increase in global average temperature did occur.[10] In addition, the sceptics argue that natural climate variability is currently poorly understood, and this variability may be greater than anticipated.[11]

Conversely, most scientists and politicians support the theory of global warming. Recent observations of rising global average air and ocean temperatures, widespread melting of snow and ice, and rising sea levels have prompted the world to acknowledge the existence of global warming. Moreover, overwhelming evidence

---

[5] NASA The Effects of Climate Change. Available at: https://climate.nasa.gov/effects/

[6] The ongoing increase of the Earth's surface temperature has been monitored over recent decades. According to IPCC, average temperatures of land and ocean surfaces have increased roughly at 0.85 [0.65 to 1.06] °C 2 over the period 1880 to 2012.

[7] IPCC above n 4.

[8] Ibid.

[9] The Climate Institute. 2010. Climate Change Making Up Your Mind. Confused by 'sceptical' claims about global warming? Here is what the science says.

[10] Leroux [6].

[11] The Climate Institute above n 9.

has also encouraged many countries to take responsibility for the key role that human activities have played in environmental degradation.

If the world continues to produce greenhouse gases into the atmosphere at present levels, global temperatures will continue to rise. The rise in temperature, in turn, will lead to the melting of polar glaciers, sea level rise, increased intensity and frequency of storms, changes in the amount and cycles of precipitation, and changes in ocean currents. In addition, global warming can lead to the spread of infectious diseases such as malaria, dengue and others, increased heat-related deaths, hunger and population displacement.[12]

The world has many opportunities to reduce its greenhouse gas emissions. A number of countries have made efforts to reduce domestic emissions. Attempts have also been made at the international level to address this problem. Nevertheless, it is clear that further progressive action is needed to stabilise greenhouse gas emissions.

1.3   The International Response to Climate Change

International efforts to combat global warming began several decades ago. In 1990, the United Nations (UN) organised negotiations for the United Nations Framework Convention on Climate Change (UNFCCC).[13] In June 1992, at the Earth Summit in Rio de Janeiro, many countries signed the UNFCCC. The parties to the Convention agreed to take measures to reduce the level of greenhouse gases in the atmosphere. The UNFCCC has set a goal for industrialised countries to limit greenhouse gas emissions to 1990 levels in 2000. The UNFCCC's goal was voluntary and merely a few countries have implemented mandatory measures to reduce greenhouse gas emissions and achieve this target.[14]

In December 1997, the third Conference of the Parties to the UNFCCC adopted the Kyoto Protocol. The Kyoto Protocol has been signed and ratified by 192 countries. The Kyoto Protocol aimed for a 5.2% reduction by industrialised countries of six GHGs (CO_2, CH_4, N_2O, SF_6, HFC and PFC) below 1990 levels by the end of 2012.[15] According to the Protocol, only Annex I[16] countries that ratified the Protocol

---

[12] IPCC above n 4.

[13] United Nations Framework Convention on Climate Change. United Nations 1992.

[14] In particular, Scandinavian and some other European countries have implemented policies to reduce GHG emissions since 1990.

[15] The Kyoto Protocol 1997. The Kyoto Protocol to the United Nations Framework Convention on Climate Change.

[16] Annex I parties are: Austria, Belgium, Bulgaria, Czech Republic, Denmark, Estonia, European Community, Finland, France, Germany, Greece, Ireland, Italy, Latvia, Liechtenstein, Lithuania, Luxembourg, Monaco, Netherlands, Portugal, Romania, Slovakia, Slovenia, Spain, Sweden, Switzerland, United Kingdom of Great Britain and Northern Ireland, United States of America, Canada, Hungary, Japan, Poland, Croatia, New Zealand, Russian Federation, Ukraine, Norway, Australia and Iceland.

had reduction obligations during the treaty's first commitment period, 2008–2012, while developing countries did not have such obligations.

1.3.1   The Kyoto Protocol

The 1997 Kyoto Protocol introduced the so-called flexible mechanisms. The implemented mechanisms are Emissions Trading (article 17), Joint Implementation (JI – article 6) and Clean Development Mechanism (CDM – article 12).[17] These three flexible mechanisms included in the Protocol aimed to help Annex I parties to achieve their emission reduction targets in 'the most cost-effective and efficient way.'[18] They introduce flexibility and facilitate compliance of parties with their respective emission reduction commitment. Article 17 of the Kyoto Protocol enables emissions trading to reduce global emissions, and states that: 'Parties included in Annex I may participate in emissions trading for the purposes of fulfilling their commitments under Article 3.'[19]

The mechanisms' operating rules were agreed at the seven's Conference of the Parties (COP), 2001, in Marrakech.[20] It was suggested that such flexible mechanisms would make strict reduction targets more acceptable for the participants.[21] In particular, the flexible mechanism:

- Stimulate sustainable development through technology transfer and investment.
- Help countries with Kyoto commitments to meet their targets by reducing emissions or removing carbon from the atmosphere in other countries in a cost-effective way.
- Encourage the private sector and developing countries to contribute to emission reduction efforts.[22]

Under the Kyoto emissions trading mechanism, Annex I Parties[23] to the Kyoto Protocol are assigned allowances (or credits), which authorise them to pollute to their committed emissions levels.[24] The Kyoto Protocol enabled the countries to buy

---

[17] The Kyoto Protocol introduced three market-based/flexible mechanisms, thereby creating what is now known as the carbon market. These are CDM, JI and Emissions Trading

[18] The Kyoto Protocol above n 15.

[19] Ibid.

[20] Ganga and Armitage [2].

[21] Grubb et al. [4].

[22] UNFCCC. Mechanisms under the Kyoto Protocol.

[23] Annex I and Annex B Countries/Parties are the signatory nations to the Kyoto Protocol that are subject to caps on their emissions of GHGs and committed to reduction targets–countries with developed economies. Annex I refers to the countries identified for reduction in the United Nations Framework Convention on Climate Change (UNFCCC) while the Annex B is an adjusted list of the countries identified under the Kyoto Protocol.

[24] UNFCCC. Kyoto Protocol Reference Manual.

emissions reduction units if they cannot otherwise meet their emissions targets, and to sell the units if they anticipate surpassing their targets. A national registry was established by each Annex I Party for monitoring and recording transfers and acquisitions of these units.[25] Businesses or non-governmental organisations and other entities were allowed to participate in trading under the supervision of their national registry. The Kyoto emissions trading is intended to be used as 'supplemental to domestic action' to achieve reduction targets.[26] However, each party was required to hold a minimum amount of emissions reduction units created and traded under the flexible mechanisms of the Kyoto Protocol.[27]

1.3.2   CDM and JI

The Joint Implementation (JI) and the Clean Development Mechanism (CDM) aimed to facilitate creation, regulation and transfers of the emissions reduction units between the Kyoto Protocol parties.[28] The JI mechanism enables Annex I countries to transfer, acquire, or finance reductions of GHG emissions, described as Emission Reduction Units (ERUs), between Annex I countries only. These ERU's may be generated by projects that reduce anthropogenic GHG emissions or enhance the removal by sinks of such emissions. The JI mechanism allows Annex I countries to invest in GHG emissions reduction projects in the other Annex I countries and receive in return emissions reduction units resulted from such projects.[29] Annex I Parties may allow private sector entities to participate, within their powers, in the production, transfer or acquisition of ERUs. The acquisition of ERUs cannot be used as a substitute for domestic actions.

The key feature of JI mechanism is that all the emissions reductions need to be verified and each Emission Reduction Unit has a serial number that is a unique identifier for the ERU. Many Annex I countries have developed ERU acquisition programs. In particular, the Netherlands actively participated in JI, purchasing emission reduction units through investments in projects in Central and Eastern Europe included in Annex I. JI is less prominent than its counterpart in developing countries, the CDM. However, JI has stimulated green investment and allowed project

---

[25] Ibid.

[26] The Kyoto Protocol 1997, Article 17.

[27] These units are: Assigned Amount Units (AAUs) are allowed emissions and were introduced under the Kyoto Protocol for the Parties of the Kyoto Protocol that have accepted targets for limiting or reducing emissions. These targets are demonstrated as levels of allowed emissions. The Emission Reduction Unit (ERU) is a unit representing one tonne of carbon dioxide of emissions issued for projects registered under the Joint Implementation mechanism. Certified Emission Reduction units (CERs) are created as a result of the CDM projects.

[28] The Kyoto Protocol 1997, Article 6, 12.

[29] UNFCCC above n 24.

emission reductions to be transferred between Annex I countries, adding an important element of flexibility to the Kyoto Protocol.

The Clean Development Mechanism is designed to be used by Annex I countries to finance GHG emissions reductions in developing countries.[30] Annex I parties receive certified emission reductions (CERs) in return for their investments. Private and public entities are authorised to participate in CDM activities. Part of the proceeds from the sales of CERs was transferred to the Adaptation Fund.[31] This fund helps developing countries, especially those vulnerable to climate change, meet the costs of adaptation.[32]

The CDM created momentum towards a global carbon market and helped resources and technology cooperation to support low-carbon development in non-Annex I countries. Annex I countries were able to reduce the cost burden in meeting their targets by taking advantage of low-cost emission reduction opportunities. The CDM also promoted flexibility when emission targets are not met by acting as a 'safety valve' for carbon prices.[33] The CDM was one of the first steps towards the development of a global carbon market, by facilitating the transition from CDM trading to international emissions trading.[34] The CDM can be interpreted as a transitional tool of international climate co-operation.

It has been argued that JI projects are more environmentally robust than the CDM because they are carried out between peer countries committed to a specific goal of reducing greenhouse gas emissions.[35] All JI parties are interested in the projects leading to current and future emission reductions. On the other hand, the CDM operated outside the established absolute emission limits. Developing host countries may be more interested in attracting investment than in following all formal procedures. Arguably the CDM created incentives to delay domestic climate policy in developing countries and limits the ambitions for low-carbon transformation in industrialised countries.[36]

A key consideration for the JI and especially CDM projects is additionality. Additionality refers to an effort that is supplementary to the business-as-usual scenario in at least two areas:

- the additionality of financial contributions of developed countries to mitigation efforts in developing countries; and
- the additionality of GHG emissions reduction generated by the projects.[37]

---

[30] The Kyoto Protocol 1997, Article 12(2).

[31] Adaptation Fund. Available at: https://www.adaptation-fund.org/about/

[32] Ibid.

[33] UNFCCC above n 24.

[34] Dellink Rob et al. [10].

[35] Lambert Schneider [12].

[36] Ibid.

[37] Streck [14].

If emission reductions in developing countries were not additional, they led to an increase of the overall cap of emissions established by commitments of the Annex I parties to the Kyoto Protocol.[38] Establishing additionality is a burdensome, time-consuming and costly process that remains challenging due to the ambiguity of its definition.[39] The absence of additionality accounted for the absolute majority of all rejections of projects submitted for registration under the CDM and JI.[40] There is also risk of fraud where projects can be financially viable on their own and that the emission reductions generated by the CDM project are not additional.[41]

The joint implementation mechanism has also received some criticism. Research by the Stockholm Environment Institute has shown that joint implementation, while designed to support climate change mitigation, seriously undermines global action to combat climate change.[42] In a random sample of 60 JI projects, 73% of offsets is generated by projects that were not additional, which means that some projects would probably have been implemented even without carbon revenues.[43]

The Kyoto Protocol has been criticised on the grounds that it only applies to developed countries, excluding developing countries, which account for more than half of global greenhouse gas emissions. Some commentators argue that the Kyoto Protocol and its flexible mechanisms did not provide a real solution for the problem of climate change.[44] Others justifiably argue that this was an essential first step in combating climate change.[45]

It is hard to deny that the emission trading introduced by the Kyoto Protocol at least provided some governments an incentive to introduce or propose national ETS. One of the factors prompting politicians in the EU, Australia, New Zealand and some other countries to consider GHG trading at the national level is the possibility of its linkage with the mechanisms of the Kyoto Protocol. According to information submitted by Parties to the Kyoto Protocol in 2014, their total greenhouse emissions at the end of the first commitment period were 22.6% lower than the base year 1990.[46] The Kyoto Protocol was therefore a success.

---

[38] Ibid.

[39] Figueres Christiana and Streck Charlotte [1].

[40] Gillenwater [3].

[41] Ibid.

[42] Kollmuss et al. [5].

[43] Ibid.

[44] McKibbin and Wilcoxen [7] and Ganga and Armitage [2].

[45] Grubb et al. [4] and Michaelowa [8].

[46] UNFCCC. Kyoto Protocol 10th Anniversary – Timely Reminder Climate Agreements Work.

1.4   Paris Agreement

The first commitment period under the Kyoto Protocol ended in 2012. There were attempts to restart operation of the Kyoto Protocol but the parties to the Protocol have failed to reach an agreement. The annual conferences of the parties to the UNFCCC have continued to negotiate an agreement that, similar to the Kyoto Protocol, would involve GHG emissions reduction. Finally, the Paris Agreement was introduced at the 21st Conference of the Parties in Paris. The agreement has been welcomed by many as a breakthrough that will translate into real cuts in greenhouse gas emissions around the world.[47] The agreement aims to limit the global average temperature rise to 2 degrees Celsius above pre-industrial levels and reduce the global average temperature rise by 1.5 degrees Celsius above pre-industrial levels.[48]

The core element of the Paris Agreement is the concept of nationally determined contributions (NDC). Article 3 of the Paris Agreement states that the parties must make an ambitious effort to achieve the goal of the Agreement.[49] It is argued that the key to the success of the Agreement is the flexibility of the NDCs.[50] The nationally defined NDC resolves previous political differences between countries over specific targets for reducing greenhouse gas emissions.[51]

The underlying belief behind the potential success of the Paris Agreement is the voluntary unification of all countries in signing the agreement.[52] Almost all countries have now signed the agreement. US President Donald Trump announced in June 2017 his intention to withdraw from the agreement.[53] This decision drew international condemnation and led to the United States being the only one who did not support the agreement. However, in 2020, the next president, Joe Biden, announced that the United States was re-joining the Paris Agreement.

The Paris Agreement stipulates that developed countries should provide financial resources to developing countries to assist in mitigation and adaptation, in support of their existing obligations under the Agreement.[54] In 2019, the Organisation for Economic Cooperation and Development (OECD) released a report stating that the climate finance provided by developed countries for climate action in developing

---

[47] Clemencon Raymond [9].

[48] The Paris Agreement, Article 2.

[49] The Paris Agreement, Article 3.

[50] Stavins and Stowe [13].

[51] Rogelj et al. [11].

[52] Stavins and Stowe above n 50.

[53] On 1 June 2017, President of the USA, Donald Trump announced the withdrawal of the USA from the Paris Agreement. However, a country must wait four years before they can officially withdraw from the Paris Agreement, the USA would only legally be removed from the PCA if declared by the President of the USA in 2020. Victoria Han. 2017. Trump's promise: Withdrawing from the Paris Climate Agreement. 29(4) Environmental Claims Journal, 337.

[54] The Paris Agreement Article 9.

countries reached USD 71.2 billion in 2017 and the Paris Agreement's goal to reach USD 100 billion in annual climate finance by 2020 is still attainable.[55]

The 2015 Paris Agreement builds on the Clean Development Mechanism and Joint Implementation to introduce the Sustainable Development Mechanism (SDM).[56] The proposed mechanism adds a more specific target to 'deliver an overall mitigation in global emissions.'[57] There is disagreement over whether and how many mitigations methodologies and carbon credits from the Kyoto era can be allowed under the Paris Agreement.[58] Countries that do not support the use of existing certified emission reduction units argue that low demand and a large number of unused units will support the already low prices for these carbon units.[59] This further reinforces fears that investors will not be motivated to fund new projects to reduce emissions under the Sustainable Development Mechanism.

Article 6.4 of the Paris Agreement states that the market-based mechanism should provide an overall reduction in global emissions. The mitigation should go beyond what would have happened if the market-based mechanism had not been applied. The CDM has facilitated the transfer of CO_2 emission reductions between countries, but the CDM has also undermined the achievement of reduction targets due to several issues, including additionality. The proposed sustainable development mechanism addresses this issue through Article 6.4, which requires emission reductions to be 'additional to any that would otherwise occur'. Furthermore, the current proposal includes various baseline tests to ensure additionality of projects. Thus, the introduction of an improved mechanism with these tests would be useful to refine the expected emission reduction targets for states and would also provide a clearer methodology for assessing additionality.

The fact that most countries have acceded to the Paris Agreement demonstrates the potential for global efforts to combat climate change. However, the enthusiasm of the parties should not automatically be perceived as countries' guarantee to reduce GHG emissions as their signature does not equal to legal obligations to achieve the targets under the Agreement.[60] The Agreement requires countries to attempt to achieve their nationally determined contributions.[61] Although unification of countries in signing the Paris Agreement appears to indicate that most countries are determined to achieve the goals of the Paris Agreement, the fact that there is no

---

[55] OECD. Climate finance for developing countries. Organisation for Economic Co-operation and Development. Paris.

[56] The Paris Agreement, Article 6.4.

[57] Carbon Market Watch. Building blocks for a robust Sustainable Development Mechanism. Policy Brief May 2017.

[58] Stavins and Stowe above n 50.

[59] Ibid.

[60] Rowe Walker [15].

[61] The Paris Agreement article 4.

legal obligation means each countries' signature is, at this stage, a non-binding declaration to reduce greenhouse gas emissions.

The flexibility of the NDCs may also be the Agreement's greatest disadvantage since the NDC is voluntarily determined by each individual country. It is also argued that even if all current NDCs are reached, global emissions in 2030 will still be higher than what is required to stop global warming.[62] Thus, while flexibility for NDCs was necessary to ensure a high level of participation, the fact that all countries can determine their own contributions undermines the ultimate goal of the Paris Agreement.

The European Union has committed the most towards the targets of the Paris Agreement compared to other nations. In 2021, the EU and its member states implemented legally binding targets to reduce 55% of GHG emissions by 2030 and an objective to achieve climate neutrality in the EU by 2050.[63] Despite, the very ambitious GHGs reduction targets the European Union's share of the worldwide GHG emissions is about 11%, thus other countries need to follow suit to ensure the success of the Paris Agreement.[64]

1.5   Summary

The individual commitments of the parties to the Paris Agreement signify the necessity of substantial efforts by the participating countries to comply with their pledges under the Agreement. Some countries already have effective domestic policies to achieve their reduction targets. Nevertheless, other countries, such as Australia, need to introduce effective national policies to reduce greenhouse gas emissions that would facilitate the fulfilment of their individual commitments under the Paris Agreement.

The next chapter focuses on a range of policy instruments utilised to mitigate climate change. The review of the fundamentals for climate policy approaches lays the foundation for further analysis of Australia's efforts to reduce greenhouse gas emissions.

---

[62] Clemencon above n 47; Kevin Anderson, John F. Broderick & Isak Stoddard. 2020. A factor of two: how the mitigation plans of 'climate progressive' nations fall far short of Paris-compliant pathways, Climate Policy.

[63] European Commission. Climate Action, 2030 climate & energy framework.

[64] Ibid.

References

1. Christiana, F., & Charlotte, S. (2009). *Enhanced financial mechanisms for post 2012 mitigation* (World Bank Policy Research Working Paper No. 5008).
2. Ganga, V., & Armitage, S. (2005). The Kyoto Protocol, carbon credit trading and their impact on energy projects in Europe and the world. *International Energy Law & Taxation Review, 4*, 73–78.
3. Gillenwater, M. (2012). *What is additionality? Part 1: A longstanding problem* (Discussion Paper No. 001). Management Institute.
4. Grubb, M., Vrolijk, C., & Brack, D. (1999). *The Kyoto Protocol: A guide and assessment.* Royal Institute of International Affairs and Earthscan.
5. Kollmuss, A., Schneider, L., & Zhezherin, V.. (2015). *Has Joint Implementation reduced GHG emissions? Lessons learned for the design of carbon market mechanisms* (SEI Working Paper No. 2015–07).
6. Leroux, M. (2005). *Global warming – Myth or reality?: The erring ways of climatology.* Springer Praxis.
7. McKibbin, W. J., & Wilcoxen, P. J. 2002. *Climate change after Kyoto: A blueprint for a realistic approach.* Brookings discussion paper.
8. Michaelowa, A. (2003). Global warming policy. *Journal of Economic Perspectives, 17*, 204–205.
9. Raymond, C. (2016). Two sides of the Paris climate agreement: Dismal failure or historic breakthrough? *Journal of Environment and Development, 25*(1), 3.
10. Rob, D., et al. (2014). Towards global carbon pricing: Direct and indirect linking of carbon markets. *OECD Journal: Economic Studies, 2013/1*.
11. Rogelj, J., den Elzen, M., Höhne, N., Fransen, T., Fekete, H., Winkler, H., Schaeffer, R., Sha, F., Riahi, K., & Meinshausen, M. (2016). Paris Agreement climate proposals need a boost to keep warming well below 2 degrees celsius. *Nature, 534*, 631, 634.
12. Schneider, L. (2007). *Is the CDM fulfilling its environmental and sustainable development objectives? An evaluation of the CDM and options for improvement.* Öko-Institut.
13. Stavins, R. N., & Stowe, R. C. (Eds.). (2016, October). *The Paris Agreement and beyond: International climate change policy post-2020.* Harvard Project on Climate Agreements.
14. Streck, C. (2010). *The concept of additionality under the UNFCCC and the Kyoto Protocol: Implications for environmental integrity and equity.* University College London.
15. Walker, R. (2016). After Paris climate agreement, what's next? *China Business Review, 1*, 1.

Chapter 2
Climate Change Mitigation

Abstract The following chapter examines major policies and mitigation mechanisms designed to reduce greenhouse gas emissions including market-based and command and control mechanisms. In the context of climate change mitigation where policy makers pursue long-term fundamental behavioural change among a large group, taxes and emissions trading arguably could be more beneficial than direct regulations or other non-price instruments. It also discusses the background, development and adverse effects associated with the introduction of market-based mechanisms to mitigate emissions.

Keywords Climate change mitigation · Economic incentives instruments · Carbon taxes · Emissions trading · EU ETS

2.1 Approaches to Climate Change Mitigation

Various countries around the world have considered numerous policy options to reduce pollution. Some of the most common policies are regulatory standards, voluntary action, taxing emissions, taxing polluting products, creating an emissions trading scheme (ETS), paying polluters to abate emissions, labelling products, educating consumers, and others. Different countries use different combinations of these approaches.

Traditionally, environmental policy is implemented primarily through the use of direct regulation.[1] This approach requires polluters to meet a certain emission target, regardless of the cost of reaching that target.[2] The direct regulation approach has been successfully applied to specific environmental issues; for example, in the United States and many European countries, direct regulation has helped improve

---

[1] Fullerton and Metcalf [27] and Harrington and Morgenstern [37].

[2] Cole and Grossman [14].

air and water quality.[3] However, the urgency of today's challenges, such as climate change, is forcing countries to seek more flexible and efficient approaches to reducing pollution, such as economic incentive instruments.[4]

Economic incentive instruments can be defined as policy instruments that use prices or other economic variables to incentivise polluters to reduce emissions. These tools were first proposed by economists and then supported by politicians. It is argued that economic incentive instruments provide price signals that can induce consumers to use fewer polluting products, thereby convincing manufacturers to produce fewer of these products.[5] Most economists give preference to economic incentive instruments.[6] In many jurisdictions, economic incentive instruments are becoming increasingly popular as environmental policy instruments. While there are many alternative economic incentives, environmental taxes and emissions trading have emerged as the two main instruments of an economic incentive policy approach.

Many countries have introduced different forms of economic incentive instruments to address various environmental issues, including climate change. In particular, policymakers in several Organisation for Economic Co-operation and Development (OECD) countries have adopted emissions trading to reduce greenhouse gas emissions. For example, in the European Union (EU), the ETS has been in effect since 2005. Environmental taxes, in turn, have been introduced in a number of countries to reduce various emissions. However, only a few countries have decided to impose an explicit carbon tax. Several EU member States, especially the Scandinavian countries, have introduced carbon and energy taxes to control greenhouse gas emissions.[7] Similar tax proposals were discussed in the mid-1990s in Australia and the United States.[8] Other countries also considered introducing a carbon tax, but ultimately most rejected this policy option and opted for emissions trading instead.

Some analysts argue that an ETS will offer participants more flexibility in attaining the reduction target than a carbon tax option.[9] Due to this and several other reasons, which are discussed in the following sections, businesses tend to favour emissions trading and oppose taxes. Governments, on the other hand, are very receptive to industry arguments for protecting economic performance from any new constraints, especially taxation.[10]

---

[3] Daugbjerg and Pedersen [18] and Freeman and Kolstad [26].

[4] Pezzey [65] and European Environmental Agency [23].

[5] Ibid.

[6] See for example: Metcalf et al. [50] and Fullerton et al. [28].

[7] European Environmental Agency [23].

[8] Trade Environment Database (TED) [81] and OECD [55].

[9] McKibbin and Wilcoxen [46].

[10] Fitzgerald [25].

Significant reductions in greenhouse gas emissions are technically and economically feasible.[11] There are many policies designed to tackle climate change, which include direct regulation as well as economic incentives. Many countries have ratified the Kyoto Protocol and the Paris Agreement and adopted various domestic regulations to mitigate climate change. However, current mitigation measures are inadequate to tackle climate change. Humanity needs to change in order for economic activity to occur within the natural environment.

2.2 Economic System and Environmental Problems

Traditional economists view economic growth as one of the main goals of economic activity.[12] Economic growth is desirable and conceivable – at least for the foreseeable future. Economic growth is associated with an eradication of poverty, higher living standards, better education, longer and healthier lives, more leisure, and an even cleaner environment.[13]

Ecological economics contends that there is a limit to economic growth. This is an inevitable conclusion based on the fact that the economy is a subsystem of a larger, limited and non-growing ecosystem.[14] This problem has been recognized for a long time. A prominent report to the Club of Rome by a group of scientists drew attention to the idea of limiting growth.[15] The report argues that exponential growth cannot be sustained indefinitely because our planet's resources are limited.

While the economy grows on a physical scale, the ecosystem does not. Consequently, as the economy grows, it becomes larger in relation to the ecosystem and eventually damages the surrounding ecosystem.[16] The principle of continuous growth is doomed to fail in a finite environment.[17] Recognising the limits to economic growth is critical to protecting the environment. A shift in economics is required to reflect quality goals and changes in the way progress is measured. Modern civilisation usually measures progress quantitatively, not qualitatively.[18] Attempts are emerging to redefine progress as more sustainable so that it fully reflects qualitative values such as wise use of resources, environmental quality and good health care.

Market failure occurs when market agents do not bear the full costs of their decisions. This situation creates so-called externalities. Externalities indicate that the

---

[11] IPCC 2014 and Garnaut [29].

[12] See, for example: Hackett [35] and Daly and Farley [17].

[13] Ibid.

[14] Meadows et al. [47].

[15] Ibid.

[16] Meadows et al. above n 14 and Daly and Farley [17].

[17] Ibid.

[18] Ibid.

price of a product does not reflect correct information about the social deficiency of that product.[19] Externalities can be positive or negative. For example, positive externalities associated with education and research play a critical role in modern economic growth theory.[20] It is generally argued that as resource use intensifies, external costs increase.[21] Since the modern economic system is based on the principle of growth, manufacturers are naturally concerned with growth and profits. Producers who can use a resource without paying the full cost tend to abuse it.[22] If a manufacturer pollutes the environment, such as air, it also affects the quality of the environment for other users. When a manufacturer does not bear the full cost of environmental pollution, the burden of the resulting inadequacy falls on the whole of society.[23]

Negative externalities are a common problem in environmental contexts. Since no one owns air or water, such resources are essentially common goods: commodities available for use and overuse by all. As a result, economic agents from individuals to businesses can use these free or undervalued resources provided by the environment without incurring the full cost. If producers bear the full costs, they will produce fewer relevant products or pass the cost onto the consumer, who in turn will consume less.[24] This problem of externalities extends to many environmental problems. Externalities contribute to ozone depletion, air and water pollution, climate change and other environmental problems.

2.3   Policy Responses

The problem of environmental externalities is clearly one of the main causes of climate change. In the first half of the twentieth century, economists suggested that because externalities are associated with crucial policy values, governments should intervene in the market to correct the problem.[25] Since then, a significant body of literature has accumulated on environmental impacts, various forms of related market and government failures, and the policy tools available to address them.

Researching policy instruments for climate change mitigation requires certain theoretical components as a framework. This section begins with an analysis of the currently prevailing policy approach, in particular the direct regulation approach. It is necessary to analyse the advantages and disadvantages of direct regulation in order to understand the further development and use of alternative approaches to

---

[19] Hackett [35].

[20] Lucas [45].

[21] Klenow and Rodríguez-Clare [43].

[22] Daly & Farley above n 16, p. 11.

[23] Jaffe et al. [40].

[24] Daly & Farley above n 16, p. 10.

[25] Pigou [68] and Baumol and Oates [6].

environmental protection. Some of the alternative policy instruments are then explored, namely economic incentive instruments, voluntary agreements and environmental education.

2.3.1 Command and Control Regulations

Direct environmental regulations, also known as command and control policies, refer to laws and regulations regarding environmental standards. Initially, governments tended to respond to environmental concerns through direct regulation, which is the usual way of addressing certain issues in various sectors of public policy.[26] The United States led the way in the 1960s in the era of modern environmental regulations.[27] Command and control regulations have become popular instrument for solving pollution problems and are arguably the most widely used tools in environmental policy.[28]

Command and control standards usually provide for general pollution levels that should not be exceeded in a geographic area, or technology standards that should be used by individual industries. Failure to comply with these regulations will usually result in sanctions. Many countries have relied heavily on command and control regulations to establish and enforce standards for equipment, processes or emissions. For example, environmental policies in countries such as the United States, China, Russia, Australia and many others have been based primarily on a command and control approach.

Economists argue that command and control policies cannot provide additional incentives to reduce pollution.[29] Command and control policies can facilitate the adoption of existing control technologies, and sometimes even contribute to the development of new technologies. To make progress in protecting the environment while maintaining economic growth, businesses need both incentive and flexibility to develop and implement innovative, efficient methods of production.[30]

Typically, command and control regulations have several flaws that undermine their ability to effectively address externalities:

1. Polluters have limited choices about how to meet standards.
2. Polluters have no incentive to explore new ways to reduce their emissions.
3. Command and control regulations are often applied uniformly across broad categories of pollutions sources, but the marginal cost will differ from source to source; therefore, emission reductions are inefficient.[31]

---

[26] Stewart [78].

[27] Johnson [41].

[28] Harrington and Morgenstern [37].

[29] Stewart [79], Stavins [76] and Sinn [74].

[30] Stewart above n 29.

[31] Ibid.

An appropriate policy response to environmental concerns is to internalise the cost of environmental damage. Environmental policies that take into account externalities promote the development of cleaner technologies by increasing the demand for equipment to reduce pollution.[32] Consequently, environmental policies, to be effective and efficient, must stimulate environmentally friendly technological change. In general, the command and control approach is preferred and in some cases relatively successful in reducing pollution. However, command and control regulations do not provide a constant price signal to facilitate research on abatement technologies or improve efficiency.[33]

2.3.2 Economic Incentive Instruments

A range of policies and instruments are available to governments to create incentives to reduce greenhouse gas emissions. Given the constraints of command and control regulations, economists and policy experts are increasingly calling for alternative policy measures based on economic incentives.

Economists believe that environmental regulations are necessary for the industrialised world.[34] While a business pays for its labour, materials and plants, it does not pay for the entire amount of pollution generated. When a business does not pay for using a resource, it has little incentive to use that resource in a reasonable manner.[35] If the business incurs the costs of the resources consumed (or at least a significant part of those costs), then there appears to be an incentive to minimise pollution. When a price attached to pollution, businesses have an incentive to reduce these costs to an optimal level that is economically viable.[36] Hence, economic incentive instruments that set a price for pollution may be more appropriate than a command-and-control approach to tackling this problem.

In the context of pollution pricing, two economic incentive instruments are often discussed, namely taxes and ETSs. Taxes can set a price for externalities; the literature defines this instrument as an effective way of internalising the cost of environmental damage.[37] Emissions trading can also set up a price on externality. Taxes and emissions trading force the polluter to pay for externalised environmental damage, which coincides with the 'polluter pays principle'.[38]

---

[32] Jaffe et al. [40].

[33] Wallace [83] and Jaffe et al. [39].

[34] Ibid.

[35] Stavins [77].

[36] Hahn and Stavins [36].

[37] Nordhaus [53] and Aldy and Stavins [1].

[38] Hahn and Stavins, above n 36.

In contrast, financial incentives (or, more generally, subsidies) can also correct externalities by paying the polluter to internalise these externalities.[39] This is contrary to the polluter pays principle. Taxes and subsidies are often seen as two sides of the same coin. Both instruments are considered as measures of economic stimulus, although there are several differences between them.[40] Taxes are by definition revenue-generating and, at worst, are fiscal neutral. While subsidies are a burden on government spending. However, the most prominent feature of this instrument is that the subsidies do not comply with the polluter pays principle.

Taxes can also be used as incentives to endorse and reward certain types of behaviour, and to encourage various activities that are beneficial to society. For example, tax incentives are used to stimulate forest conservation in a number of countries, including Australia.

2.3.3 Tax Incentives for Forest Conservation in Australia

Financial incentives through the tax system are vital for the conservation of land and forests and can be an effective mechanism for the conservation of private land.[41] There are different types of taxation provisions that can be used to encourage conservation of land and forest. Australia's tax regime provides certain incentives for conservation of land and forests. However, some commentators argue that the Australian tax system was built to promote productive land use.[42] Arguably, the tax system does not adequately encourage conservation, especially private conservation,[43] despite the fact that significant part of Australia's native forests are privately owned.[44]

In Australia, as in several other industrialised countries, a substantial proportion of native forests, about 67%, are privately managed, which creates an additional level of complexity in terms of forest management and protection.[45] Generally, land clearing in Australia mostly occurs on private land.[46] Australian governments have used various mechanisms to encourage conservation on private land such as tax incentives although, the effectiveness of these incentives and their ability to support the conservation of Australia's lands and forests are questionable.

---

[39] The Institute for European Environmental Policy (IEEP) [80].
[40] European Environmental Agency [23].
[41] Ibid.
[42] Binning and Young [9] and Douglas [19].
[43] Ibid.
[44] Rawlings et al. [69].
[45] Department of Agriculture and Water Resources, Australia's State of the Forests Report. 2018. Available at: https://www.agriculture.gov.au/abares/forestsaustralia/sofr
[46] Megan [48].

In many cases, private landowners are unable to generate substantial income from their land if the land is used for conservation purposes[47] but they have to bear various ancillary costs such as rates and taxes, fire management and others.[48] Current tax regime provides limited financial assistance to those landowners who preserve their land for future generations.[49] Tax deductability is one of the key considerations that may encourage conservation. The Australian tax regime contains a number of expense deduction provisions. Section 8-1 of the Income Tax Assessment Act 1997 (ITAA97) is a general deduction provision that allows certain expenses to be deducted from assessable income. Australia's tax regime focuses on income related expenses, which means that a general deduction is only possible if the expense is incurred in producing assessable income or in carrying on a business. Accordingly, land conservation-related expenditure is not tax deductible unless it is directly linked to the commercial use of that land.[50]

Landholders carrying on a business on their land can deduct the cost of rates and land tax from their assessable income. This provides an incentive for landholders to use the land for production purposes rather than conserve it. Under Australian tax law a business is a commercial activity carried out for the purpose of making a profit or generating income. Some conservation activities may include generating income through financial incentives provided by various conservation schemes,[51] but in general, land conservation is unlikely to be income-generating or business-related and is more likely to be qualified as private or domestic in nature. Therefore, the general deduction section 8-1 does not apply to the conservation of land or forests.

ITAA97, contains many specific deduction provisions some of which relate to environmental activities. Section 40.755 provides for a deduction of expenditure incurred for environmental protection activities related to pollution and/or waste.[52] The section stipulates that environmental protection activities are carried on to prevent, fight or remedy pollution or waste: 'resulting, or likely to result, from your earning activity...'[53] It is clear from the wording of this section that the focus is on 'earning activity' and therefore land or forest conservation activities are not covered in this section.

The Australian tax system also provides deductions or concessions for landcare operations.[54] The landholder may be eligible to claim capital expenditures in relation to expenses incurred on landcare operations if the land is used for primary

---

[47] Binning and Young above n 42.

[48] Ibid.

[49] Vanclay et al. [82].

[50] See generally, Income Tax Assessment Act 1997 (Cth) s 8-1.

[51] See for example, Department of Agriculture, Water and the Environment. Incentive Programs around Australia. Available at: http://www.environment.gov.au/biodiversity/conservation/incentive-programs-around-australia

[52] Income Tax Assessment Act 1997 (Cth) s 40.755.

[53] Ibid.

[54] Income Tax Assessment Act 1997 (Cth) Subdiv. 40-G.

production or other business.[55] Deductions are available to landholders who use their land to conduct business for the 'purpose of producing assessable income'[56] and primary production.[57] However, landcare deductions provisions generally do not cover the expenditures of landholders managing permanently protected land without any income.

A possible solution to this issue for the Australian system would be to expand the scope of the landcare operations provision to include landholders involved in conservation activities. In particular, section 40.630 of ITAA97 could encompass 'conservation of environmentally sensitive land'. The expenditure of landholders involved in conservation should be deductible from their earned assessable income, regardless of the source of income. In addition, landholders involved in conservation activities should be exempted from the requirement to conduct business on the land for taxable purposes. Further, section 40.635 of ITAA97 describing the 'meaning of landcare operations' should be expanded to include 'management and restoration of environmentally sensitive land.'

One of the most relevant deductions provisions associated with conservation covenants or some other permanent protection instruments registered on title is Division 31 of ITAA97.[58] A conservation covenant can be defined as a voluntary agreement between a landholder and a designated body (usually a Covenant Scheme Provider)[59] that aims to protect the natural, cultural and/or scientific values of the land. The landholder owns, uses and lives on the land but the natural values of an area are preserved by the landholder with assistance of eligible Covenant Scheme Provider.[60]

Eligible taxpayers holding a conservation covenant could be allowed a deduction under Division 31. The deductible amount is the difference between the market value of the land prior to entering into the covenant agreement and the market value of the land immediately thereafter. A deduction is only possible if the value of the land has decreased by more than $5000 as a result of the covenant agreement[61] and if no fee has been received for entering into the covenant agreement.[62] Consequently, the deduction is not available where a landholder enters into a covenant and also participates in the National Landcare Program which provides certain grants,[63] or

---

[55] Income Tax Assessment Act 1997 (Cth) Subdiv. 40.630.

[56] Income Tax Assessment Act 1997 (Cth) s 40.25(7)(a).

[57] Income Tax Assessment Act 1997 (Cth) s 40.630(1)(b).

[58] Income Tax Assessment Act 1997 (Cth) Div. 31.

[59] Covenant Scheme Providers could be not-for-profit organisations, government agencies or local councils (listed on Register of Environmental Organisations) that can sign conservation covenants with landholders to protect land with conservation values.

[60] Covenant Scheme Providers are listed on Register of Environmental Organisations. Available at: http://www.environment.gov.au/about-us/business/tax/register-environmental-organisations

[61] Income Tax Assessment Act 1997 (Cth) ss 31.5(2)(c)-(d).

[62] Income Tax Assessment Act 1997 (Cth) s 31.5(2)(b).

[63] See National Landcare Programme. Available at: http://www.nrm.gov.au/national-landcare-programme

receives some income as part of native vegetation offsets transactions.[64] Deductions can be apportioned over 5 years to preserve tax benefits if a landholder's income for 1 year is less than the value of the land covered by the covenant agreement.[65] However, a deduction under this provision cannot result in an income tax loss.

One of the major weaknesses of the Australian tax system is that in some cases it fails to recognise the public interest character of private conservation. If the covenant occurs solely as a gift, it is accepted as a transaction in the public interest and thus is covered by applicable tax benefits, such as deductions under Section 31 of ITAA97. However, if the landowner receives some remuneration in exchange for entering into a conservation covenant agreement, the Australian tax regime classifies such transactions as private in nature and no tax benefit apply. A necessary step for the Australian tax system would be to recognise and accommodate both the public and private interest character of environmentally beneficial conservation transactions.

The Australian experience, where the deductions provided under Division 31 have been a key feature of the system for many years, suggests that such an approach per se is not enough to ensure adequate conservation incentives. It needs to be a tax incentives system which can sufficiently facilitate other characteristics, to ensure that those willing to participate in land conservation have appropriate incentive and trust in the outcomes of the process. To that end, the system must ensure a certain degree of simplicity and certainty, it must be able to provide consistent support and incentive to biodiversity conservation, it must provide flexibility in its approach and it must be effective in the sense of being beneficial and transparent for taxpayers.

Generally, the Australian tax system facilitates some incentives to encourage private land conservation but there are also a number of impediments distracting conservation. One of the possible solutions could be an introduction of a provision that focuses exclusively on land conservation. The Australian tax system has similar stand-alone provisions dealing with specific issues. For example, part 3-50 of ITAA97 institutes the system for treatment of emissions units. This approach establishes an important component of consistency and comparative simplicity in the tax treatment of emissions units. An analogous stand-alone provision covering land and forest conservation would be beneficial. This would enable the creation of an integrated tax regime for the conservation of land and related management activities. The experience of some other countries indicates that the tax system could provide more effective incentives for private conservation than those existing in Australia.[66]

There are various suitable strategies that can help promote conservation of land and forests including direct regulations, financial incentives, environmental education and others.[67] These instruments are likely to work more effectively together

---

[64] See e.g. Department of Environment, Land, Water and Planning (DELWP). Native Vegetation. Available at: https://www.environment.vic.gov.au/native-vegetation/native-vegetation

[65] Income Tax Assessment Act 1997 (Cth) Subdiv. 30-DB.

[66] Guglyuvatyy [34].

[67] OECD [57].

facilitating the change in landowners' behaviour and providing the necessary incentives for conservation.[68]

2.3.4 Environmental Education

The modern economic system needs to be changed in order to progress toward a cleaner environment, and society must make substantial progress toward environmentally responsible behaviour. It is critical to educate public to comprehend the environmental problems and potential solutions through the spread of environmental education.

Environmental education is a potentially very effective method of protecting the environment. Several countries have initiated environmental education programs in schools to change people's attitudes towards the environment and make them more environmentally conscious. For example, Italy became the first country to include climate change and sustainable development issues in its national curriculum in 2019. Numerous universities in different countries now have various programs with an environmental component. There are many environmental non-governmental organisations (NGOs) that educate the public on environmental issues in order to raise environmental awareness.

Education is essential for solving environmental problems, especially in the long term.[69] Yet, people's habits and attitudes do not change overnight, and not all polluting products can be simply replaced. Given that climate change requires an urgent change in attitudes and behavior, education as the central policy to reduce greenhouse gas emissions may not be appropriate. While environmental education is crucial as a long-term policy, it will not provide the immediate change in consumers' behaviour needed to tackle climate change.

The discussed policies have been used to mitigate climate change but only some of them can provide an accurate price for externalities.[70] Some of the non-pricing instruments allow for an implied price on greenhouse gas emissions. For example, stringent emission standards encourage manufacturers to invest in technology to meet standards, and in some scenarios the costs associated with such standards can be extremely high. Thus, direct regulation can impose implicit externalities on pollutants in certain settings.

An effective price signal can reveal significant potential for reducing greenhouse gas emissions. It is argued that in order to stop climate change, it is necessary to implement policies that set a strong price signal.[71] Economic incentive instruments,

---

[68] Natural Resource Management Ministerial Council, Australia's Biodiversity Conservation Strategy 2010–2030, Australian Government, Department of Sustainability, Environment, Water, Population and Communities, Canberra.

[69] Riethmuller and Buttriss [70].

[70] Stavins [77].

[71] Ross [71] and IPCC. AR5 Synthesis Report: Climate Change 2014.

taxes and emissions trading, are designed to correct externalities and provide an explicit price signal. In particular, taxes and emissions trading are becoming much more politically attractive instruments as they force polluters to pay for the produced pollution.

Direct regulation, financial incentives and environmental education are essential to complement policies that set prices for greenhouse gas emissions.[72] Conversely, proposing any of these measures as the main policy for reducing greenhouse gas emissions is a dubious strategy. These tools cannot provide a clear price for carbon and are therefore unlikely to facilitate the rapid changes in business and consumer behaviour required to stop climate change. Given the vast nature of greenhouse gas emissions from various sources, policies that set a clear price for carbon emissions are essential as the centrepiece of an effective climate change mitigation strategy. Accordingly, the following sections focus on two environmental policy instruments that can provide the necessary price signal, namely taxes and emissions trading.

2.4 Environmental Taxes

Taxation is one of the most powerful public policy instruments that can influence people's behaviour and economic activity.[73] Taxes can be used to deter or encourage various types of economic activity, production or consumption of various types of goods. Taxation, especially in the area of environmental policy, is recognised as the main policy instrument capable of addressing environmental policy, taxation is recognised as a major policy instrument which is capable of addressing numerous environmental issues.

The theoretical foundations of pollution taxation are attributed to the economists Pigou (1932), Coase (1960), and Baumol and Oates (1971).[74] Pigou, in his book the Economics of Welfare, recommended an emissions fee levied on polluters according to their actual output of pollution. It is one of the earliest modern concepts of environmental taxation. Pigou's general prescription for taxing externalities coincides with the polluter pays principle and is based on the notion that external costs can be internalized by imposing a tax, thereby reducing externalities.[75]

In 1960, Coase sharply criticised the Pigouvian doctrine. In his renowned article, 'The Problem of Social Cost', Coase challenged the Pigouvian tax system and argued that, under certain conditions, the market mechanism will lead to optimal resource allocation, correcting itself where externalities exist.[76] Despite this criticism, taxes are generally viewed as suitable instrument for reducing pollution.

---

[72] Ibid.

[73] Baumol and Oates [7] and Bovenberg and De Mooij [10].

[74] Pigou [68], Coase [13] and Baumol and Oates [6].

[75] Ibid.

[76] Baumol [5].

However, current concepts and practices related to environmental taxes differ from the ideal Pigouvian tax.

The central economic assumption for use of taxes in environmental policy is to transfer the costs of emissions and other externalities into the prices of goods and services. However, the notion of environmental taxation has evolved over time. There is no universally accepted definition of an environmental tax. Terminology varies from author to author and can be confusing. Therefore, for clarity and consistency in this book, the term 'environmental taxes' will encompass all pollution, emissions, green or eco-taxes. In the context of climate change, taxes on greenhouse gases are often referred to as carbon or CO_2 taxes, and this book will use the term carbon taxes.

The ideal environmental tax is described in the literature as a tax designed to incorporate external costs into prices in order to bring social and private costs closer together.[77] The most significant difficulty in assessing external costs are the lack of information and scientific uncertainty in the relationship between emissions and environmental impacts.[78] In practice, environmental taxes are often the result of a compromise between conflicting goals and lobbying groups.[79] Governments often choose to impose taxes that reflect only a small portion of environmental impacts.[80] Environmental goals are determined by a political process, and the tax rate is set at a level that should lead to the achievement of these goals, and rather than matching external costs. However, even if the environmental tax does not cover all external costs, it is still beneficial because it covers at least some negative externalities and makes polluters pay.

Environmental taxes encourage producers and consumers to be more environmentally responsible. While some environmental taxes may target consumers, others target producers. Eventually, the costs associated with tax are passed on to consumers of products or services that have caused harm to the environment.[81] Environmental taxes provide a constant incentive for polluters to reduce emissions.[82] Since the tax is paid on a per unit of emission basis, it creates an incentive for the polluter to cut emissions and increase profits.

Revenue raising function of environmental taxes is potentially the most attractive for governments. From an environmental point of view, this function is especially valuable if the revenues raised is utilised to protect the environment. In this way, the public will be confident that such tax revenue will essentially benefit the environment.[83] The revenue can also be used to lower or eliminate other taxes. For example, revenues from an environmental tax can offset taxes on labor, capital and savings,

---

[77] Andersen [3].

[78] Newell and Pizer [52].

[79] Bressers and Huitema [11] and Grattan Institute [30].

[80] Ibid.

[81] Creedy and Sleeman [16].

[82] Fischer et al. [24].

[83] EEA [20].

potentially resulting in double dividends.[84] The double dividend hypothesis is often associated with revenue generating environmental policy instruments. The double dividend is potentially beneficial for the environment and the economy.

In 1972, the OECD adopted the concept of raising taxes on 'bads' (e.g. pollution) and lowering taxes on 'goods' (e.g. labour).[85] Peirce later developed the double dividend theory as applied to environmental taxation.[86] The double dividend emerges, on the one hand, in solving the environmental problem (an externality) and, on the other hand, in reducing the excess burden from other taxes. Revenue generating economic incentive instruments can contribute significantly to labour and capital tax reform.[87] There are supporters and opponents of the double dividend theory. Nonetheless, a cleaner environment rather than a possible double dividend should be the main argument in favour of environmental taxes.

2.4.1 Taxes and Climate Change

In the climate change mitigation context, when policymakers pursue long-term fundamental change in the behaviour of a large group, taxes appear to be more beneficial than direct regulations. This is due to the fact that taxes can address the problem of reducing greenhouse gas emissions, set a price on emissions and induce society to change their behaviour and become environmentally responsible.

Unlike direct regulations that require all polluters to cut their emissions by the same amount without considering their costs, a carbon tax gives the polluter two options: pay the tax or reduce pollution.[88] Thus, polluters who face high costs of reducing emissions will pay the tax, while those who face low costs of reducing emissions will reduce their emissions instead. All management decisions are in the hands of the polluter, and results will be achieved without costly government monitoring. Consequently, a given level of overall reduction in greenhouse gas emissions can be achieved with a tax more efficiently than through regulation.[89]

The tax system is effective in delivering price signals at the systemic level, which is essential to target large-scale sources of greenhouse gas emissions.[90] Changes in the tax system affect the economic decision-making process by all market participants, from corporate leaders to individual consumers and politicians. For example, if a tax raises the price of fossil fuels to reflect externalities from fossil fuel use, the change will affect all economic decisions related to energy. Changing fossil fuel

---

[84] Parry [63].

[85] OECD [54].

[86] Pearce [64].

[87] Ibid.

[88] Fullerton et al. [28].

[89] Ibid.

[90] Pearce above n 86, p. 942.

prices through the tax system is also administratively straightforward as it removes
the need to disseminate information about the measure – the signal is contained
directly in the market price. Since most individuals and legal entities are already
filing tax returns, the new tax instrument does not require a new interaction between
the taxpayer and the government.

2.5   Emissions Trading

Emissions trading, or tradable permit systems, is another actively discussed economic incentive instrument designed to address various environmental issues.
Similar to taxes, emissions trading is intended to set a price for pollution, leading to
cost-effective emissions reduction. Tradable permit systems have been used to
tackle a variety of environmental issues such as air pollution, fishing, water management, waste management and land use since the 1970s.[91] Emissions trading has also
been considered as a potential instrument for reducing greenhouse gas emissions
since the early 1990s.

The original concept of emissions trading could be attributed to the renowned
economist Coase.[92] Baumol and Oates develop Coase's theory of property rights.[93]
They assume that property rights should be distributed among authorised businesses, thereby giving them the right to pollute at a certain level.[94] Governments will
grant as many pollution rights as necessary to indicate the overall allowable pollution level. Businesses that keep their emissions below the specified level can trade
the remaining pollution rights to those who need to emit more than they are
allowed to.[95]

This mechanism gives everyone a constant incentive to reduce pollution and
develop and use new technologies to protect the environment, thereby contributing
to the efficient use of resources.[96] The use of tradable permits for pollution control
has evolved from academic theories to a central element of environmental policy in
several countries. It has also become an important element of the UN program to
mitigate global warming and, in particular, the Kyoto Protocol.

Probably the most important aspect of the ETS is its design. There are various
options for the design of this tool; however, it seems that the most acceptable ETS
schemes are cap and trade along with baseline and credit. Under a cap and trade
scheme, the government first sets emission reduction targets and then, based on that

---

[91] EPA [22].

[92] Coase above n 74.

[93] Baumol and Oates [7].

[94] Ibid.

[95] Ibid.

[96] Ibid.

figure, sets a cap on total emissions.[97] The government then issues a fixed number of tradable permits to participants, where each permit representing a specific authorisation to pollute (for example, one ton over a specific period). Participants are closely monitored by the regulatory body and at the end of the compliance period, participants must have sufficient number of permits to cover their pollution for this period. Participants anticipating a shortfall in permits at the end of the compliance period can purchase additional permissions from members with excess permits.

According to the alternative baseline and credit approach (also called rate-based), emission allowances are determined according to some business parameter, such as energy consumption.[98] The set of allowable emissions for the respective years determines the baseline, which depends on the performance of the participant. The business will receive credits if emissions are less than the baseline for a given year. Credits can be sold to another business that cannot reach the baseline. Unlike cap and trade, there is no general cap in the baseline and credit scheme; rather, allowable emissions increase and decrease with economic activity. The baseline and credit system does not guarantee that a specific reduction target will be achieved. Consequently, if the achievement of the reduction target is critical, the cap and trade is likely to prevail.

2.5.1   Methods of Emission Permits Allocation

The cap and trade scheme appears to have evolved into the mainstream policy. One of the most important aspects of a cap and trade is the initial allocation of emission allowances. Typically, permits are either sold at auction or issued free of charge to emitters, and in some cases governments may use both of these methods.

The choice of initial permits allocation can have a large impact on the overall cost of implementing an ETS. Many of the existing ETSs use an initial allocation of emission allowances based on specific historical emissions data.[99] When based on historical emission levels, the free allocation method is commonly referred to as 'grandfathering'. This method minimises the political debate associated with the transition from command and control to ETS. It is also attractive to businesses as the existing pollution sources receive very valuable permits free of charge.[100] Under the grandfathering method, existing polluters may be in even better position than they were with the command and control system. Free permits are more likely to generate the necessary political support from the respective industries.

---

[97] EPA above n 91.

[98] Ibid.

[99] OECD [58].

[100] Ibid.

There is generally a prevalent argument in the literature for auctioning, also known as revenue generation permit allocation.[101] Grandfathered permits have adverse consequences in the same way as taxes, but unlike taxes, permits are deprived of the potential benefit of revenue raising. There are very few ETSs that distribute permits on a purely auction basis. More often than not, a small percentage of permits remain with the government and are auctioned off, while most permits are issued free of charge.[102] For example, this method was initially used in the European Union Emissions Trading System (EU ETS).[103]

A hybrid approach involving auctions and free allocation may provide some of the benefits of the former, along with some of the characteristics of the political acceptability of the latter.[104] The auction price reflects the cost of using the environment and corrects market imperfection. Auctioning a significant portion of emission permits will generate revenue for the government that can be used not only to reduce distorting taxes, but also to finance various environmental projects.

Either the environmental tax or the ETS provide a continuing incentive for technological innovation and the development of new efficient abatement methods, helping to further reduce emissions and thus reducing the tax or permit costs for polluters. In the same way as environmental taxes, ETS provides a clear price signal and incentives for the development of cleaner technologies. There are also adverse effects associated with environmental taxes and emissions trading.

2.6 Adverse Effects of Taxes and ETS

While economists are concerned with efficiency, policy makers tend to focus on the adverse effects associated with the use of economic incentive instruments.[105] The likelihood that carbon taxes or ETS could affect the competitiveness of a country, sectors and regions, or specific firms is being seriously considered by governments.[106]

One of the main arguments of policy makers in justifying resistance to economic stimulus instruments is the fear of a decline in competitiveness in the most affected sectors of the economy.[107] Whether taxes or emissions trading are introduced, there are negative implications for the international competitiveness of certain industrial sectors or companies if such instruments are applied in a non-global manner.[108]

---

[101] Freeman and Kolstad [26], Ellerman and Joskow [21] and OECD above n 132.

[102] OECD above n 99.

[103] EU ETS is discussed in more details below.

[104] Aldy et al. [2].

[105] Zhang and Baranzini [86].

[106] Guglyuvatyy [33] and Saddler et al. [72].

[107] OECD [59].

[108] Ibid.

According to the OECD, concerns about international competitiveness are the reason why proposals to enact in 1993 the thermal tax in the United States, a Greenhouse Levy in Australia in 1994 and Council Directive to establish a common EU agenda for energy taxation in 2003, were discarded.[109]

In the short term, higher energy prices could potentially have a detrimental effect on the competitiveness of certain sectors, but labour-intensive sectors often benefit from effective economic incentive instruments.[110] A recent study demonstrates that the EU ETS has had a limited negative impact on competitiveness due to over-allocation of emission allowances, which has led to falling prices and the ability of companies in some sectors to pass on costs to consumers.[111] The study assumes that in Phases III and IV (2013–2030) the negative impact on competitiveness from the EU ETS will be minimal or not at all.[112] Furthermore, the EU ETS is said to have significantly stimulated innovation that is greater than its cost effects.[113] Thus, it can be concluded that more attention is paid to competitiveness issues than is justified.

Another potential negative impact of carbon taxes and ETS relates to income distribution. Low-income households tend to spend proportionately more on electricity or fuel than wealthier groups.[114] Consequently, they are relatively more vulnerable to higher prices on fossil fuels than high-income households. Normally, households can change their behaviour in response to taxes and, accordingly, reduce the use of resources. Consequently, the burden of economic instruments is often exaggerated.[115] A recent study examines the potential impact of the carbon price in Europe.[116] The study found that if the distributional effects of price increases associated with carbon pricing are properly addressed, the negative impacts can be reduced. The study confirms that if the revenues from carbon pricing are returned to households, the incomes of the poorest households will rise, reducing the regressive effect of the carbon price. Competitiveness and distribution concerns are believed to arise if carbon taxes or emissions trading are imposed. However, these concerns can be effectively eliminated.

---

[109] OECD [56].

[110] EEA [20].

[111] Joltreau and Sommerfeld [42].

[112] Ibid.

[113] Lin et al. [44].

[114] Callan et al. [12] and Owen and Barrett [62].

[115] Schlegelmilch [73].

[116] Montenegro et al. [51].

2.7 Carbon Taxes in Practice

Numerous countries have introduced various environmental taxes, fees and charges.[117] Some are effective, others are nominal and provide minimal benefits.[118] In the context of climate change, several countries have introduced carbon taxes. There have also been attempts to introduce carbon taxes internationally, namely for EU member States.[119]

In 1991, the European Parliament formally proposed a 50/50 tax on CO_2 and energy in response to concerns about climate change.[120] However, EU members were unable to agree on uniform carbon taxes.[121] The opposition in Europe stems in part from the perception that the introduction of carbon taxes will affect the international competitiveness of industries.[122] Nevertheless, European countries were the first to introduce taxes aimed at reducing greenhouse gas emissions. The first carbon taxes in Europe were introduced in the Nordic countries and the Netherlands.[123] Since then, many countries have followed suit. The carbon tax is currently imposed in various countries around the world, including European nations, Canada, Singapore, Japan, Argentina and others.[124]

Carbon taxes are levied on various types of greenhouse gas emissions such as carbon dioxide, methane, nitrous oxide and fluorinated gases. The amount and scope of the carbon taxes varies from country to country. For example, in Spain, the carbon tax applies only to fluorinated gases, covering only 3% of the country's total greenhouse gas emissions. Norway, by contrast, has removed most of the exemptions and reduced rates, as a result more than 60% of its emissions are covered by a carbon tax.

Sweden introduced a carbon tax in 1991 as a supplementary tax to the existing structure of energy taxes.[125] Since 1991, the Swedish tax system had been modified several times. The primary political objective of this programme was to shift the tax burden away from taxes levied on labour and to compensate the loss in revenues by increasing environmental taxes.

Carbon tax is paid in full by households and service sector companies, and by manufacturing companies, except for fuels used for heating. The Swedish carbon

---

[117] For example, many of the EU and OECD countries have introduced different environmentally related taxes (detailed data could be found at the EEA/OECD Economic Instruments database http://www2.oecd.org/ecoinst/queries/index.htm).

[118] OECD [56].

[119] Trade Environment Database [81].

[120] Ibid.

[121] Ibid.

[122] Zhang and Baranzini [86].

[123] Finland was the first country to introduce a carbon tax in 1990. Carbon and Energy Taxes (Finland), UCD Dublin.

[124] OECD [61].

[125] Speck et al. [75].

taxes generally imply a high burden for fossil fuel energy consumption on industry however, the impact of the Swedish carbon tax on energy and resource efficiency of industries has been limited due to exemptions and refunds to industries.[126] Nonetheless, the Swedish experience is one of the most successful examples where the current carbon tax is €112.08 per tonne of CO_2.[127] Greenhouse gas emissions in Sweden have decreased by 25% since 1995, while the country's economy grew by 75% over the same period.[128]

Political realities and special political interests (eg, concerns about competitiveness) are evident in many exceptions for a number of industries in several countries.[129] Carbon taxes in many countries do not provide a uniform coverage of greenhouse gas emissions from different sources. Instead, they provide a wide range of different rates, exemptions, and reimbursements for the industries covered. Carbon taxes can be ineffective if the right price signal is not transmitted throughout the economy. Taxes need to be accurately designed to address the issue effectively.

2.8   Emissions Trading in Practice

The first ETSs were introduced in the United States in the 1970s. Emissions trading originally focused on a variety of environmental or resource issues such as air pollution, water pollution, fishing, waste management and land use.[130] The earliest example related to air pollution was the SO_2 trading scheme introduced in the United States in 1995 as a result of the 1990 Clean Air Act amendments.[131]

Emissions trading has been seen as a promising instrument for reducing greenhouse gas emissions since the early 1990s. In 1997, the US proposed emissions trading as one of the Kyoto Protocol mechanisms. At a later stage in the negotiations, emissions trading was implemented in the Kyoto agreement, despite objections from many parties.[132]

Despite its attractiveness for parties of the regime, the emissions trading mechanism adopted in the Kyoto Protocol had a crucial drawback. Creating an international GHG trading system is fully justifiable but such an imperative issue as 'hot air' is left unanswered.[133] Attributable to the economic decline and end of the government subsidies to industry in Eastern Europe and of the former Soviet Union,

---

[126] Ibid.
[127] OECD. Taxing Energy Use 2019: Figure 3.7, Explicit carbon taxes do not cover all energy-related emissions.
[128] Ibid.
[129] Ibid.
[130] Anderson [4].
[131] Ibid.
[132] Philibert and Reinaud [66].
[133] Grubb et al. [31] and Woerdman [84].

emissions levels in this region have fallen significantly since 1990 and stayed well below their 1990 level until 2012.[134] The CO_2 emissions of the Russian Federation and the Ukraine, two of the largest polluters, have decreased by 36.9% and 55.3% between 1990 and 1998 respectively.[135] Since these countries did not accept reduction targets in line with this decrease, a large amount of tradable emissions is available in the system. This 'hot air' surplus, if it were to be transferred to another country to help in achieving its reduction commitments, would make emissions higher than in the absence of trading because what is being traded does not represent a real reduction.[136]

Russia's decision to ratify the Protocol in 2004 led to a surge in carbon credits trading in the EU. A critical aspect of the hot air problem is that the wealthiest nations could have benefited from such emissions trading arrangements.[137] Furthermore, if some countries make profits from the sale of hot air, it created a stimulus for developing countries to join the treaty with the intention of profiting from the sale of generous emissions assignments.[138] Also, the integrity of the Kyoto Protocol's goal might have been lost because some Annex I Parties bought up emissions credits elsewhere, thereby severely weakening the climate change regime by failing to take any domestic action.[139] Overall, it is hard to deny that the emissions trading introduced by the Kyoto Protocol provided some governments an incentive to consider a national ETS.

2.8.1 European Union Emissions Trading Scheme

In the early 2000s, emissions trading to reduce greenhouse gas emissions was introduced in Denmark, followed by the EU and the UK.[140] The EU ETS was established in 2003 and entered into force on 1 January 2005. The EU ETS was introduced primarily to help member States meet their emission reduction targets under the Kyoto Protocol.[141]

The first phase (2005–2007) of the EU ETS, during which the cap was set at a relatively low level, was described as a trial period.[142] European Commission Directive 2003/87/EC created a legislative mandate for the member States to quantify, allocate and reduce greenhouse gas emissions with the goal of reducing at least

---

[134] Woerdman [84].

[135] Grubb et al. above n 133.

[136] Yamin [85].

[137] Grubb et al. above n 133.

[138] Woerdman above n 134.

[139] Ibid.

[140] Philibert and Reinaud [67].

[141] Convery and Redmond [15].

[142] Ellerman and Joskow [21].

5% of total greenhouse gas emissions, including carbon dioxide, compared to 1990 levels.[143] The EU ETS initially covered the greenhouse gas emissions of around 12,000 companies in the oil and gas, power, pulp and paper, cement, glass and steel sectors across the EU.

The Emissions Trading Directive required each member State to formulate a National Allocation Plan (NAP) for greenhouse gas emissions in accordance with the parameters set by the Directive.[144] During Phase I and Phase II, each member State determines the total quantity of permits needed and a specific allocation plan (in consultation with the public and industries). These proposed National Distribution Plans were then submitted to the European Commission for review and approval. The European Linking Directive provides European companies the opportunity to invest in emission reduction projects in developing countries and convey carbon credits for use in the EU ETS.[145] Credits created under the CDM or JI mechanisms of the Kyoto Protocol were recognised as equivalent to EU ETS's allowances.

Phases I and II were accompanied by serious legal problems with respect to the NAPs of the member States in the EU courts. During that period, environmental litigation constituted a significant and growing proportion of the casework of the European Union courts.[146] The litigation mainly concerned the failure of the member States to comply with the deadline for the enactment of domestic legislation in accordance with the Directive or, after the adoption of such legislation, the failure to notify the Commission of such enactment.[147]

For the first commitment phase, the member States have allocated 95% of the permits free of charge, with the option to auction the remaining 5% of permits.[148] During the first phase of the EU ETS, countries lobbied for the interests of their industries and, as a result, issued too many permits.[149] Consequently, the economic and environmental performance of the EU ETS has been undermined by widespread free allocation of emission permits and the fact that some industries have been able to shift EU ETS-related costs to households through higher prices for related goods such as electricity. Notably, the EU ETS does not address the distribution concerns, leaving it to the discretion of the member States. Several EU governments have tried

---

[143] Directive 2003/87/EC of the European Parliament and of the Council of 13 October 2003 establishing a scheme for greenhouse gas emission allowance trading within the Community and amending Council Directive 96/61/EC 2003.

[144] Ibid.

[145] Ibid.

[146] Court of Justice of the European Union: Annual Report 2009.

[147] See for example, Case C-107/05 Commission of the European Communities v Republic of Finland [2005] OJ C93/22, C60/10 14; Case C-122/05 Commission of the European Communities v Republic of Italy [2006] OJ C115/15, C165/10.

[148] For the period 2005–2007 only one member country, Denmark, chose to auction the maximum percentage allowed by the EU Emissions Trading Directive.

[149] Ellerman and Joskow above n 142.

to tackle this problem. For example, Spain allows retail price increases for certain groups of customers, but only up to a certain limit.[150]

However, the first phase of the EU ETS was a trial period aimed at providing countries with the experience they need to successfully implement emissions trading. Phase 1 of the EU ETS scheme was used to ascertain carbon market pricing methodologies and to establish methods for monitoring, reporting and verifying emissions. Given the experimental nature of Phase I, only 0.13% of the permits were sold at auction, with most of the permits distributed free of charge to enterprises through the National Distribution Plan (NAP) system.[151]

Some observers also criticised the EU ETS because of its limited scope.[152] Another widespread criticism relates to low permit prices, which discourage polluters from reducing greenhouse gas emissions and investing in abatement technologies.[153] Despite this criticism, a recent study shows that between 2008 and 2016, CO_2 emissions fell by about 3.8%. These reductions represent almost half of the EU Kyoto target.[154] Consequently, even the low carbon price set under the EU ETS provides a signal that governments will impose costs on greenhouse gas emissions in the long term, and this signal leads to a noticeable reduction in greenhouse gas emissions.[155]

In the second phase (2008–2012), the EU ETS set much more ambitious emission reduction targets that improved its efficiency. The cap was about 13% lower than the first period and 6% lower than the comparable 2005 emission level.[156] In Phase II, the auctioning received greater use. EU countries were allowed to auction up to 10% of permits and more States use this option. Nonetheless, only a small number of countries opted to use the auctioning option and none of them use the maximum percentage allowed.[157] The number of sectors covered by the ETS has increased. Fewer quotas were issued, free quotas were reduced by 10%, and the penalty for non-compliance was increased from €40 to €100 per tonne of CO_2.[158] The national registers have been replaced by a single European Union register for stationary installations.

Since 2013, the EU ETS has expanded to include non-CO_2 greenhouse gases and all major industrial emission sources. Phase III led to the creation of a more centralised system to relieve pressure on governments to calculate and monitor the allocation and compliance of ETS quotas. Phase III also introduced an aviation emissions trading scheme on all flights operating in EU airspace. Airlines are required to

---

[150] Hourcade et al. [38].

[151] Bayer and Aklin [8].

[152] Metcalf and Weisbach [49].

[153] OECD [60].

[154] Bayer and Aklin [8].

[155] Ibid.

[156] Ellerman and Joskow above n 142.

[157] Ibid.

[158] Grubb et al. [32].

surrender a portion of their allowances for every ton of CO_2 emitted during operation.[159] This requirement applies not only to European airlines, but also to any airlines operating in or out of the EU. However, the inclusion of international aviation was legally challenging for the EU.

The lack of clear reporting prior to the introduction of the EU ETS means that it was difficult to attribute the emission reductions achieved by many Member States directly to the ETS program. Between 1990 and 2004, emissions in the EU decreased by 0.08% per year. Since the implementation of the ETS program in 2005, emissions have decreased at a rate of 1% per year between 2005 and 2010.[160] From 2013 to 2020, the EU ETS annual cap decreased by 1.74% per year, and from 2021 it will decline by 2.2% per year.[161] According to the European Commission, in 2020 emissions in the sectors covered by the ETS will be 21% lower than in 2005, and in 2030 emissions in the covered sectors will be 43% lower than in 2005.[162]

Overall, EU ETS sets clear targets and limits for businesses, eliminating uncertainty and confusion in the process. Tightening these limits will allow the ETS to reach the reduction target. The EU entered Phase IV of the scheme in 2021, which will last until 2030. To accelerate the rate of emission reductions, the total number of emission permits will be reduced annually by 2.2% from 2021, up from 1.74% at present.[163] The EU ETS has resulted in some declines in greenhouse gas emissions, although more significant efforts are required to achieve long-term emissions reduction targets in Europe.

2.9   Summary

Climate change is one of the most challenging issues in the world today, requiring effective policies to reduce greenhouse gas emissions. Taxes and emissions trading have a number of advantages over non-pricing policy instruments. The main rationale for highlighting these measures is that they incorporate the polluter pays principle and can play a central role in effective climate change mitigation policies. Taxes and emissions trading can equally set an immediate price for carbon, thereby creating an incentive for ongoing efforts to reduce greenhouse gas emissions.

The era of stimulating growth through increased production and overuse of natural resources must be clearly reconsidered. Traditional policies and recommendations may have been relevant in the past, but the current climate change challenge requires urgent action. Many countries have implemented various strategies to reduce greenhouse gas emissions and to tackle climate change. The experience of

---

[159] Ibid.

[160] Ibid.

[161] European Commission. EU Emissions Trading System.

[162] Ibid.

[163] Ibid.

these countries suggests that, despite some shortcomings, certain policy instruments can help reduce greenhouse gas emissions. Despite these successful examples in the implementation of effective climate change policies, some countries, including Australia, are still reluctant to follow suit. The next chapter discusses the current approach to climate change policy in Australia.

References

1. Aldy, J. E., & Stavins, R. N. (2008). *Economic incentives in a new climate agreement*. The Harvard Project on International Climate Agreements.
2. Aldy, J. E., Krupnick, A. J., Newell, R. G., Parry, I., & Pizer, W. (2009). *Designing climate mitigation policy*. Resources for the Future, Discussion Paper 08-16.
3. Andersen, M. (1994). *Governance by green taxes: Making pollution prevention pay*. Manchester University Press.
4. Anderson, R. C. (2001). *The United States experience with economic incentives for protecting the environment*. EPA, EE-0216B.
5. Baumol, W. J. (1972). On taxation and the control of externalities. *American Economic Review, 62*, 307–322.
6. Baumol, W. J., & Oates, W. E. (1971). The use of standards and prices for protection of the environment. *The Swedish Journal of Economics, 73*, 42–54.
7. Baumol, W. J., & Oates, W. E. (1988). *The theory of environmental policy*. Cambridge University Press.
8. Bayer, P., & Aklin, M. (2020). The European Union emissions trading system reduced CO_2 emissions despite low prices. *PNAS, 117*(16), 8804–8812.
9. Binning, C., & Young, M. (1999). *Talking to the taxman about nature conservation: Proposals for the introduction of tax incentives for the protection of high conservation value native vegetation*. National R&D Program on Rehabilitation, Management and Conservation of Remnant Vegetation, Research Report 4/99, Environment Australia.
10. Bovenberg, A. L., & De Mooij, R. A. (1994). Environmental levies and distortionary taxation. *American Economic Review, 84*, 1085–1089.
11. Bressers, H., & Huitema, D. (1999). Economic instruments for environmental protection: Can we trust the "magic carpet"? *International Political Science Review, 20*, 175–196.
12. Callan, T., Lyons, S., Scott, S., Tol, R., & Verde, S. (2008). *The distributional implications of a carbon tax in Ireland* (Working paper 250). Economic and Social Research Institute.
13. Coase, R. H. (1960). The problem of social cost. *Journal of Law and Economics, 3*, 1–44.
14. Cole, D. H., & Grossman, P. Z. (1999). When is command-and-control efficient? Institutions, technology and the comparative efficiency of alternative regulatory regimes for environmental protection. *Wisconsin Law Review, 5*, 887–938.
15. Convery, F., & Redmond, L. (2007). Market and price developments in the European Union Emissions Trading Scheme. *Review of Environmental Economics and Policy, 1*, 88–111.
16. Creedy, J., & Sleeman, C. (2006). Carbon taxation, prices and welfare in New Zealand. *Ecological Economics, 57*, 333–345.
17. Daly, H. E., & Farley, J. (2004). *Ecological economics: Principles and applications*. Island Press.
18. Daugbjerg, C., & Pedersen, A. B. (2004). New policy ideas and old policy networks: Implementing green taxation in Scandinavia. *Journal of Public Policy, 24*, 219–249.
19. Douglas, R. (2002). *Potential effects of selected taxation provisions on the environment. Consultancy report to The Productivity Commission*. Productivity Commission.
20. EEA. (2005). *Market-based instruments for environmental policy in Europe*. Technical report 8.

21. Ellerman, D., & Joskow, P. L. (2008). *The European Union's emissions trading system in perspective*. Pew Centre on Global Climate Change.
22. EPA. (2008). *Clean air markets-allowance trading*. United States Environmental Protection Agency.
23. European Environmental Agency. (2006). *Using the market for cost effective environmental policy*. EEA report 1/2006.
24. Fischer, C., Parry, I., & Pizer, W. (2003). Instrument choice for environmental protection when technological innovation is endogenous. *Journal of Environmental Economics and Management, 45*, 523–545.
25. Fitzgerald, J. (2006). *Lobbying in Australia: You can't expect anything to change if you don't speak up*. Rosenberg Publishing.
26. Freeman, J., & Kolstad, C. (2007). *Moving to markets: Lessons from twenty years of experience*. Oxford University Press.
27. Fullerton, D., & Metcalf, G. E. (2001). Environmental controls, scarcity rents, and pre-existing distortions. *Journal of Public Economics, 80*, 249–267.
28. Fullerton, D., Leicester, A. & Smith, S. 2008. Environmental taxes. National Bureau of Economic Research, Working Paper 14197.
29. Garnaut, R. (2008). *Garnaut climate change review*. Cambridge University Press.
30. Grattan Institute. (2018). *Who's in the room? Access and influence in Australian politics*. Report No. 2018-12.
31. Grubb, M., Vrolijk, C., & Brack, D. (1999). *The Kyoto Protocol: A guide and assessment*. Royal Institute of International Affairs and Earthscan.
32. Grubb, M., Laing, T., Sato, M., & Comberti, C. (2012). *Analyses of the effectiveness of trading in EU-ETS*. Climate Strategies.
33. Guglyuvatyy, E. (2012). Australia's carbon policy – A Retreat from core principles? *eJournal of Tax Research, 10*(3), 552–572.
34. Guglyuvatyy, E. (2018). Failing to see the wood for the trees? A critical analysis of Australia's tax provisions for land and forest conservation. *Australian Tax Forum, 33*(3), 551–571.
35. Hackett, S. C. (1998). *Environmental and natural resources economics: Theory, policy, and the sustainable society*. M.E. Sharpe.
36. Hahn, R., & Stavins, R. (1991). Incentive-based environmental regulation: A new era from an old idea. *Ecology Law Quarterly, 18*, 1–42.
37. Harrington, W., & Morgenstern, R. D. (2004). *Economic incentives versus command and control* (p. 13). Resources for the Future 152.
38. Hourcade, J. C., Demailly, D., Neuhoff, K., Sato, S., Grubb, M., Matthes, F., & Graichen, V. (2007). *Climate strategies report: Differentiation and dynamics of EU ETS industrial competitiveness impacts*. CWPE 0712.
39. Jaffe, A., Newell, R. G., & Stavins, R. N. (2002). Environmental policy and technological change. *Environmental and Resource Economics, 22*, 51–70.
40. Jaffe, A., Newell, R. G., & Stavins, R. N. (2005). A tale of two market failures: Technology and environmental policy. *Ecological Economics, 54*, 164–174.
41. Johnson, S. M. (2004). *Economics equity and the environment*. Environmental Law Institute.
42. Joltreau, E., & Sommerfeld, K. (2019). Why does emissions trading under the EU Emissions Trading System (ETS) not affect firms' competitiveness? Empirical findings from the literature. *Climate Policy, 19*(4), 453–471.
43. Klenow, P. J., & Rodríguez-Clare, K. (2005). Externalities and growth. In P. Aghion & S. Durlauf (Eds.), *Handbook of economic growth*. Elsevier.
44. Lin, W., Chen, J., Zheng, Y., & Dai, Y. (2019). Effects of the EU Emission Trading Scheme on the international competitiveness of pulp-and-paper industry. *Forest Policy and Economics, 109*.
45. Lucas, R. E. (2002). *Lectures on economic growth*. Harvard University Press.
46. McKibbin, W. J. & Wilcoxen, P. J. 2006. A credible foundation for Long Term International Cooperation on Climate Change. ANU, CAMA Working Paper Series.

47. Meadows, D. H., Randers, J., Meadows, D. L., & Behrens, W. W. (1974). *The limits to growth: A report for the Club of Rome's project on the predicament of mankind.* Universe Books.
48. Megan, E. (2016). Deforestation in Australia: Drivers. *Trends and Policy Responses, 22*((2) PCB 130), 130–150.
49. Metcalf, G., & Weisbach, D. (2009). The design of a carbon tax. *Harvard Environmental Law Review, 33*, 500–556.
50. Metcalf, G. E., Palstev, S., Reilly, J., Jacoby, H., & Holak, J. (2008). *Analysis of U.S. greenhouse gas tax proposals.* MIT Joint Program on the Science and Policy of Global Change.
51. Montenegro, R. C., Lekavicius, V., Brajkovic, J., Fahl, U., & Hufendiek, K. (2019). Long-term distributional impacts of European cap-and-trade climate policies: A CGE multi-regional analysis. *Sustainability, 11*(23), 6868.
52. Newell, R. G., & Pizer, W. A. (2003). Regulating stock externalities under uncertainty. *Journal of Environmental Economics and Management, 45*, 416–432.
53. Nordhaus, W. D. (2007). To tax or not to tax: Alternative approaches to slowing global warming. *Review of Environmental Economics and Policy, 1*, 26–44.
54. OECD. (1972). *Recommendation of the council on guiding principles concerning international economic aspects of environmental policies.* Organization of Economic Cooperation and Development, Council Document C(72)128.
55. OECD. (2001). *Environmentally related taxes in OECD countries: Issues and strategies.* OECD.
56. OECD. (2006). *The political economy of environmentally related taxes.* OECD Publishing.
57. OECD. (2007). *Instrument mixes for environmental policy.* OECD Publishing.
58. OECD. (2008). *Environmentally related taxes and tradable permit systems in practice.* Centre for Tax Policy and Administration.
59. OECD. (2010). *Taxation, innovation and the environment.* OECD publishing.
60. OECD. (2018). *Effective carbon rates 2018: Pricing carbon emissions through taxes and emissions trading.* OECD Publishing.
61. OECD. (2019). *Taxing energy use 2019: Using taxes for climate action.* OECD Publishing.
62. Owen, A., & Barrett, J. (2020). *Reducing inequality resulting from UK low-carbon policy.* Climate Policy.
63. Parry, I. W. (1995). Pollution taxes and revenue recycling. *Journal of Environmental Economics and Management, 29*, 564–577.
64. Pearce, D. (1991). The role of carbon taxes in adjusting to global warming. *Economic Journal, 101*, 938–948.
65. Pezzey, J. (2003). Emission taxes and tradeable permits – A comparison of views on long-run efficiency. *Environmental and Resource Economics, 26*, 329–342.
66. Philibert, C., & Reinaud, J. (2004). *Emissions trading: Taking stock and looking forward.* OECD/IEA.
67. Philibert, C., & Reinaud, J. (2007). *Emissions trading: Trends and prospects.* OECD.
68. Pigou, A. C. (1932). *The economics of welfare.* Macmillan.
69. Rawlings, K., Freudenberger, D., & Carr, D. (2010). *A guide to managing box gum grassy woodlands.* Department of the Environment, Water, Heritage and the Arts.
70. Riethmuller, S. H., & Buttriss, G. J. (2008). *Closing the gap between pro-environmental attitudes and behaviour in Australia.* Australian National University.
71. Ross, G. (2008). *The Garnaut climate change review final report.* Cambridge University Press.
72. Saddler, H., Muller, F., & Cuevas, C. (2006). *Competitiveness and carbon pricing border adjustments for greenhouse policies.* The Australia Institute Discussion Paper 86.
73. Schlegelmilch, K. (2000). Ecological tax reform. In H. Ott & T. Takeuchi (Eds.), *Towards coordinated climate protection strategies.* Wuppertal Spezial.
74. Sinn, H. W. (2007). *Public policies against global warming* (CESifo Working Paper 2087). Munich.

75. Speck, S., Andersen, M. C., Nielsen, H., Ryelund, A., & Smith, C. (2006). *The use of economic instruments in Nordic and Baltic environmental policy 2001–2005*. National Environmental Research Institute.
76. Stavins, R. (1997). *Policy instruments for climate change: How can national governments address a global problem?* Resources for the Future Discussion Paper 97-11.
77. Stavins, R. (2001). *Experience with market based environmental policy instruments*. Resources for the Future 01-58.
78. Stewart, R. (1985). Economics, environment, and the limits of legal control. *Harvard Environmental Law Review, 9*, 1–22.
79. Stewart, R. (1996). United States environmental regulation: A failing paradigm. *Journal of Law & Commerce, 15*, 585–596.
80. The Institute for European Environmental Policy (IEEP). (2012, October). *Study supporting the phasing out of environmental harmful subsidies*. Final Report.
81. Trade Environment Database (TED). (1997). *TED case studies: EC carbon tax*. TED.
82. Vanclay, J., Thompson, D., Sayer, J., McNeely, J., Kaimowitz, D., Gibbs, A., Crompton, H., Cameron, D., & Bevege, I. (2006). *A proposal for stewardship support to private native forests in NSW*. The Southern Cross Group of forest researchers and practitioners.
83. Wallace, D. (1995). *Environmental policy and industrial innovation: Strategies in Europe, the U.S.A., and Japan, London*. Earthscan Publications.
84. Woerdman, E. (2005). Hot air trading under the Kyoto Protocol: An environmental problem or not? *European Environmental Law Review, 14*, 71–77, p. 73.
85. Yamin, F. (1998). The Kyoto Protocol, origins, assessment and future challenges. *Review of European Community and International Environmental Law, 7*, 113–127.
86. Zhang, Z., & Baranzini, A. (2004). What do we know about carbon taxes? An inquiry into their impacts on competitiveness and distribution of income. *Energy Policy, 32*, 507–518.

Chapter 3
Australia and Climate Change

Abstract This chapter examines the evolution of climate change law and policy in Australia. In particular, the focus of this chapter is on historical development of the climate regime in Australia. Key developments and mechanisms related to climate change mitigation are reviewed, including greenhouse gas reporting arrangements, renewable energy policy and other regulations aimed at reducing greenhouse gas emissions. The Emissions Reduction Fund and other mechanisms, which are central to Australia's current climate policy, are also critically reviewed.

Keywords Evolution of climate policies · Australia · Carbon pricing · Emission reduction fund

3.1 Australia and the Climate Change Agenda

Australia has a history of constantly changing climate change initiatives and policies. A number of measures to reduce greenhouse gas emissions in Australia have been on the Commonwealth and State agendas over the past few decades. Successive Australian governments have been committed to the introduction of either a carbon tax or an emissions trading to mitigate climate change. Some of Australia's policies to reduce greenhouse gas emissions have been successfully implemented, some have been introduced and then repealed, and some have never reached the implementation stage.

In June 1992, at the Earth Summit in Rio de Janeiro, many nations joined the United Nations Framework Convention on Climate Change (UNFCCC).[1] Australia actively participated in the 1992 Rio Earth Summit, supporting the Summit's goals, which were shaped by the pursuit of sustainable development. Australia also signed

---

[1] UN. Sustainable Development Knowledge Platform. Agenda 21, Available at: https://sustainabledevelopment.un.org/outcomedocuments/agenda21

© The Author(s), under exclusive license to Springer Nature Singapore Pte Ltd. 2022
E. Guglyuvatyy, *Climate Change, Forests and Federalism*,
https://doi.org/10.1007/978-981-19-0742-5_3

the UNFCCC and later the Kyoto Protocol[2] supporting the reduction of greenhouse gas emissions.

Climate change has been on the agenda in Australia before the Rio summit. In 1989, ahead of the 1990 Federal elections, major political parties discussed the introduction of policies to reduce Australia's greenhouse gas emissions. The Labor Party discussed the goal of reducing greenhouse gas emissions by 20% by 2005.[3] Concurrently, the Liberal Party pursued similar policies and claimed to be ahead of Labor on climate change and many other environmental concerns during the 1990 election campaign.[4] In the 1990s, interest in the issue of climate change dissipated. Environmental issues gradually disappeared from the political agenda and did not figure in the 1993 election campaign. The recession of the 1990s, the rise of neoliberalism among Labor ministers, and the dominance of the energy and coal lobbies were among the factors that led to the decline of the environmental agenda.[5]

Despite declining attention to environmental issues in the early 1990s, a number of measures to reduce greenhouse gas emissions in Australia have been on the Commonwealth and State agendas in recent decades. Successive Australian governments have pledged to introduce either a carbon tax or an Emissions Trading Scheme (ETS) designed to reduce greenhouse gas emissions.[6] Several Australian States and local governments have implemented pollution and waste disposal charges. There is some experience with ETS in Australia. At the State level, New South Wales introduced the Greenhouse Gas Abatement Scheme (GGAS) in 1997. It was one of the world's first mandatory GHG ETS.

3.1.1   Greenhouse Gas Abatement Scheme

The scheme, developed as a baseline and credit scheme, aimed to reduce per capita greenhouse gas emissions associated with electricity consumption in New South Wales from 8.65 tonnes of CO_2 in 2003 to 7.27 tonnes of CO_2 by 2007 and contain this level until 2020.[7] The scheme has imposed obligations to reduce greenhouse gas emissions on electricity retailers, producers and some large energy consumers in New South Wales.[8] Liable entities could achieve their targets either directly or by acquiring New South Wales Greenhouse Gas Acceptance Certificates (NGACs),

[2] Australia signed the Kyoto Protocol on April 29, 1998, but ratified it on December 12, 2007. The Kyoto Protocol entered into force in Australia 90 days after ratification on March 11, 2008.

[3] Staples [15].

[4] Ibid.

[5] Ibid.

[6] Wilder and Fitz-Gerald [17].

[7] IPART, NSW Greenhouse Gas Reduction Scheme Strengths, weaknesses and lessons learned Greenhouse Gas Reduction Scheme. July 2013.

[8] MacGill et al. [9].

which represent one tonne of avoided greenhouse gas emissions from activities that reduce or offset emissions.

The scheme has endorsed trading in NSW Greenhouse Gas Emission Reduction (NGAC) certifications, that can be used to offset emissions associated with electricity purchases. The certificates can be generated by accredited Abatement Certificate Providers (ACPs) who have taken action to reduce emissions from existing generators, generated electricity using low-emission technologies, improved energy efficiency, sequestered carbon in forests, and reduced emissions from industrial processes. The overall purpose of the scheme was to show that the market mechanism can achieve environmental objectives at a relatively low cost to consumers and government by providing financial incentives to businesses participating in the scheme.

The scheme's emission reduction targets (also known as benchmarks) increased in each of the first four years of its existence - but they remained at 2007 levels until the scheme was ended in 2012.[9] The benchmarks were expressed in tonnes of carbon dioxide equivalent (tCO_2e) per capita, making it difficult to calculate individual abatement certification obligations for benchmark participants. Participants generally found it difficult to predict their future compliance obligations with an acceptable level of confidence.[10]

The baseline and credit design of the scheme was built on complex and imputed emission reductions that required counterfactual assumptions that would otherwise have occurred. There was no general emission limit, but instead accredited projects were allowed to create NGACs where each certificate corresponds to abated emissions, for example 1 tonne of CO_2.[11] The absence of emissions cannot be measured but must be estimated based on a prediction of what would happen without the scheme. A transparent approach that includes measuring absolute emission reductions would be more effective.[12] The scheme also did not set a price for greenhouse gas emissions, but rather set a price for NGAC, which effectively means a reduction in imputed emissions from the estimated baseline. Pricing greenhouse gas emissions is a vital function of any emissions trading scheme that aimed for emission reductions and the transition to a carbon-neutral economy.

The additionality issue is inherent in the baseline and credit scheme. Additionality is a critical factor from a climate change perspective. The GGAS did not properly address the additionality issue. Cap and trade emissions trading schemes are based on physical measurable emissions rather than abstract concepts of emission reductions based on an estimate of what would otherwise have happened. To eliminate the additionality problem, many countries, including the European Union, have instead adopted cap and trade schemes. A baseline and credit approach adopted by the Clean Development Mechanism projects implies rigorous and formal tests for

---

[9] IPART, above n 7.

[10] MacGill et al, above n 8.

[11] Ibid.

[12] Ibid.

additionality. The CDM requires projects pursuing accreditation to clearly demonstrate that they would reduce emissions in addition to business as usual. The NSW GGAS did not include additionality testing, thereby reducing the efficiency and effectiveness of the scheme.

Some amendments have been made to the GGAS to address additionality and some other related issues. For example, from July 1, 2010, pre-existing projects were generally excluded from the scheme.[13] Specifically, pre-GGAS projects could only generate emission reduction certificates when they exceeded a non-zero production baseline that reflected their pre-GGAS production.

Since its launch in 2003, the scheme has helped cut greenhouse gas emissions by about 16 million tonnes, according to the NSW government.[14] However, it appears that at least 83% in 2003, 76% in 2004 and 52% in 2005 of NGACs were created by preexisting low-emission plants that did not need to increase their production to create GGAS certificates.[15] Several of the projects that led to the creation of NGAC continue to raise serious concerns. Undoubtedly, some of the projects implemented by NGAC represent additional emission reductions and the scheme has resulted in certain additional investment in low emission generators. Also, when the scheme closed on June 30, 2012, most participants were well prepared for the transition to Australia's national carbon pricing mechanism, which commenced on July 1, 2012. Overall, the GGAS projects have provided some reduction in greenhouse gas emissions and created an additional financial incentive for enterprises to improve their energy efficiency.

3.2   Climate Change Mitigation 1990s–Early 2000s

The Australian Commonwealth Government has supported the national ETS after years of support from State governments. The government of Prime Minister John Howard has taken climate change into consideration since its election in 1996[16] In 1997, Prime Minister Howard revealed a $180 million package to reduce greenhouse gas emissions. The Australian Greenhouse Gas Authority was created in 1998 with a $555 million package to develop measurement and monitoring systems for greenhouse gas emissions and energy efficiency standards for a range of devices and equipment.[17]

---

[13] IPART, above n 7.

[14] MacGill et al, above n 8.

[15] Passey et al. [13].

[16] Senator Birmingham speech in the Australian senate Thursday, 20 September 2007. Available at: http://parlinfo.aph.gov.au/parlInfo/search/display/display.w3p;query=Id%3A%22chamber%2Fha nsards%2F2007-09-20%2F0379%22

[17] Anita Talberg, Simeon Hui and Kate Loynes. Australian climate change policy to 2015: a chronology. Parliament of Australia.

In 2000, Howard's government implemented the Renewable Energy (Electricity) Act 2000, which established the Mandatory Renewable Energy Target Scheme to stimulate renewable energy development across Australia.[18] The Mandatory Renewable Energy Target imposed an obligation on electricity retailers and large consumers to deliver an additional 2% (or 9500 GWh) of their electricity from renewable sources by 2010. The scheme has prompted the development of renewable energy and helped to reduce greenhouse gas emissions from electricity generation. The scheme initially stimulated demand for electricity generated from renewable sources. However, in January 2004, a review of the Mandatory Renewable Energy Target (Gambling Report) indicated that the scheme would not be able to stimulate investment in renewable energy after 2007.[19]

Despite international pressure, particularly from the European Union, the Howard government declared that Australia would not endorse the Kyoto Protocol.[20] The rationale for this was the following: the treaty did not cover 70% of global emissions; developing countries did not have emissions reduction target; and the largest source of greenhouse gases at the time, the United States, did not ratify the treaty. However, Howard's government announced that Australia would accomplish its Kyoto target without ratification.[21]

A government energy report published in 2004 reaffirmed its decision not to ratify the Kyoto Protocol, abandon emissions trading, and not to impose mandatory renewable energy target.[22] Two years later, in 2006, climate change mitigation agenda received strong political momentum in Australia.[23] The Labor opposition, in particular, called for the ratification of the Kyoto Protocol, demonstrating a deeper commitment to combating climate change.[24] Howard's government responded by accepting a recommendation from a joint business-government working group to implement an emission trading.[25] Furthermore, the government announced that Australia would support a new international agreement to curb the growth of greenhouse gas emissions, provided that it binds all countries.[26]

In late 2006, Prime Minister Howard announced that Australia would introduce domestic ETS[27] no later than 2012.[28] A Ministerial Emissions Trading Task Force

---

[18] Ibid.

[19] Recommendations of the Tambling report. Available at: https://www.aph.gov.au/Parliamentary_Business/Committees/Senate/Environment_and_Communications/Completed_inquiries/2004-07/renewableenergy/report/e02

[20] Bamsey and Rowley [1].

[21] Staples above n 3.

[22] Zahar et al. [18].

[23] Howard [7].

[24] Staples above n 3.

[25] Howard, above n 23.

[26] Staples above n 3.

[27] McKibbin [10].

[28] Ibid. Note, however, that in 2005, the Australian State and Territories issued a discussion paper concerning a national emissions trading scheme which would cover the power generation sector.

was established to develop the ETS. The Task Force stated that 'the most efficient and effective way to manage risk is through market mechanisms' and therefore ETS is the preferred emission reduction mechanism for Australia.[29] This statement based on the vision that it is much better to let the market 'decide which new or existing technologies will reduce emissions at the lowest cost' rather than leave it to the government to 'pick the winners'.[30] The main features of the ETS recommended by the Task Force were based on the 'cap and trade' system.

The Task Force also noticed that ETS should be flexible, technology impartial and operate at a national level.[31] The report recommended the introduction of additional measures to address market failures not corrected by the ETS, including informational, educational and voluntary strategies, as well as subsidies for the development of new technologies.[32] As a first step towards emissions trading, a National Greenhouse Gas and Energy Reporting Bill was introduced in September 2007. This bill created an emissions reporting system that covers about 75% of all GHG emissions in Australia.[33] The scheme covered transport and other fuels, as well as all six greenhouse gases under the Kyoto Protocol. The reporting requirements include about 700 Australian companies and provide a single national reporting system that eliminates duplicate mechanisms developed by State and territory governments.[34]

Overall, the Howard government began taking action to combat climate change shortly after the 1996 elections. A number of programs initiated by the government have laid the foundation for further efforts to reduce greenhouse gas emissions. Howard's government has managed to introduce some of the most advanced climate change policies, such as the creation of the world's first national agency to address greenhouse gas emissions. However, some of these policies were introduced as a result of a political compromise rather than a strong desire by the government to reduce greenhouse gas emissions.[35] It is also considered that the government's actions on environmental issues were rather disordered. In fact, it appears that Prime Minister Howard did not have a specific vision of the environment, and as a result, his government's actions on climate change were not very effective.[36]

[29] Prime Ministerial Task Group on Emissions Trading, Report of the task group on emissions trading, 2007, Commonwealth Australia Available at https://apo.org.au/sites/default/files/resource-files/2007-06/apo-nid968.pdf

[30] Ibid.

[31] Ibid.

[32] Ibid.

[33] Netbalance. 2012. National Greenhouse and Energy Reporting. Department of Sustainability, Environment, Water, Population and Communities.

[34] Ibid.

[35] Guglyuvatyy and Stoianoff [6].

[36] Ibid.

3.3 Climate Change Mitigation 2007–2013

Ahead of the 2007 Federal elections, opposition leader Kevin Rudd announced Labor's commitment to tackle climate change issues.[37] In particular, Labor declared their intention to ratify the Kyoto Protocol, supported the Garnaut Review.[38] They also reaffirmed the introduction of emissions trading, which included specific targets aligned with a 60% reduction in greenhouse gas emissions below 2000 levels by 2050.[39]

The Rudd government ratified the Kyoto Protocol and announced the Australian Carbon Reduction Scheme (ACPRS) during its first year in office. The proposed ACPRS was intended to help meet Australia's emission reduction target and support a global response to climate change.[40] In July 2008, the Rudd government issued a Green Paper outlining the preferred design for the ACPRS and identifying issues for further consideration.[41] Relevant legislation that included programs for adaptation to the unavoidable impacts of climate change was proposed in May 2009.[42]

The proposed legislation established a series of short-term annual total emission caps from 2011, in line with the long-term goal of reducing Australia's emissions by 60% by 2050.[43] The emission caps should be gradually reduced over time, which should lead to higher greenhouse gas prices, increased investment in low-emission technologies and lower total emissions. Participants in the scheme will need to have sufficient emission permits to account for their annual emissions. Permits will be awarded through auctions, although some will be handed out free of charge to help emissions-intensive trade exposed (EITE) industries adapt to a carbon constrained economy.[44]

The Rudd government has pledged to cut emissions by 5% by 2020 from 2000 levels.[45] This level of reduction was expected to rise to 25% of the reduction in greenhouse gas emissions by 2020, subject to an international agreement.[46] The Rudd government postponed the introduction of the scheme (first 2010, then 2011) until 2012 due to pressure from the industry.[47] The compromise also included a

---

[37] Gartrell [4].

[38] The Garnaut Climate Change Review, was commissioned in 2007 and the final report was released on 30 September 2008.

[39] Ibid.

[40] Maraseni Tek Narayan et al. [11].

[41] Guglyuvatyy and Stoianoff above note 35.

[42] Ibid.

[43] Maraseni et al., above n 40.

[44] Ibid.

[45] Beder [2].

[46] Maraseni et al., above n 40.

[47] Beder above n 45.

relatively low fixed starting price of permits of $10 per tonne, and even more permits had to be handed out free of charge.[48]

The proposed ACPRS included the same GHGs as the Kyoto Protocol - emissions from stationary energy, transport, fugitive emissions, industrial processes, and waste.[49] The ACPRS was expected to cover about 1000 major emission sources, which are estimated to account for 75% of Australia's emissions.[50] The proposed emissions trading scheme was broader in scope than any other ETS proposed or operating in Australia or overseas, including the EU ETS and the 2006 States and Territory proposal.

The proposed legislation provided for unlimited linkage, that is, participants could use an unlimited number of Kyoto units[51] in addition to the ACPRS permits.[52] The unrestricted use of Kyoto units could have potentially compromised the achievement of Australia's national emission reduction target. It would have been prudent for the Rudd government to set a limit on the number of the Kyoto units that can be surrendered by companies, as stipulated in the EU ETS. This easing of restrictions demonstrated the need to tackle the tense political environment surrounding the attempt to introduce the ACPRS.

There are other examples where the government needed to placate various industries, such as emissions intensive industries. The Rudd government has offered to allocate over 30% of annual permits free of charge to help emissions intensive industries meet their ACPRS obligations. However, the level of assistance to the emissions intensive sectors was expected to decrease over time.[53] The proposed legislation specified that coal fired generators was to receive additional assistance, such as support for the development of carbon capture and storage technologies, as well as essential assistance for affected workers, communities and regions.[54] The coal fired electricity producers were also expected to receive direct cash payments.[55] Considering that more than 70% of Australia's electricity at that stage was generated from coal combustion, any major climate change solution must target the electricity sector directly.

The greenhouse gas price through the ACPRS was intended to stimulate the energy sector to reduce the intensity of greenhouse gas emissions. However, the substantial compensation to emissions intensive industries and coal fired electricity generators offered by the then Rudd government would have almost certainly send the wrong message to business and distort the ultimate goal of climate change

---

[48] Maraseni et al., above n 40.

[49] Ibid.

[50] Ibid.

[51] Kyoto units are units created using the mechanisms of the Kyoto Protocol.

[52] See, Carbon Pollution Reduction Scheme Bill 2010. Available at: https://www.aph.gov.au/Parliamentary_Business/Bills_Legislation/Bills_Search_Results/Result?bId=r4281

[53] Lyster Rosemary [14].

[54] Ibid.

[55] Ibid.

mitigation. Substantial compensation can negate, or at least seriously reduce, perceived incentives to develop new energy efficient manufacturing processes or switch to renewable energy sources. Moreover, if producers do not take into account the cost of emissions, the environmental impact of such a climate change mitigation policy will be eliminated.

The proposed assistance to the energy intensive industries and the energy generators would subsidise the largest carbon sources, leaving the burden of ACPRS to less polluting industries and diminishing the opportunity for cleaner energy sources.[56] Rather than compensating certain businesses, ACPRS revenues could be better leveraged by offering additional assistance to introduce low emission technologies. To eliminate the adverse impact of ACPRS on the households' cost of living, the Rudd government has planned increasing payments to people receiving social benefits beyond automatic indexation.[57] Furthermore, targeted support was to be provided to low-income households through the tax and social security systems.[58]

Similar to the European emission trading system, the Rudd Government has underlined the international harmonisation of emissions trading. The ETS creates a unit of exchange for harmonisation: permits, expressed in terms of greenhouse gas emissions units. Hence, if it is cheaper to cut emissions in Indonesia than in Australia, the first step is to cut emissions where the costs are lower.[59] The Rudd government has also offered free permits for vulnerable industries.[60] Unlike the EU ETS, the ACPRS addressed distributional issues associated with emission trading.[61] Despite numerous concessions in the ACPRS design, industry pressure led the Rudd government to postpone the launch of the proposed scheme (initially 2010, then 2011) until 2012.[62] The ACPRS legislation was defeated twice in the Australian parliament in 2009, and as a result, the Rudd government suspended ACPRS in April 2010.[63]

The Rudd government's ACPRS scheme was a significant improvement over the Howard government's approach to climate change policy. The ACPRS legislation took into account the work of Prime Minister Howard's Emissions Trading Task Force, the National Emissions Trading Task Force, Garnaut Climate Change Review and Treasury Modeling.[64] This methodological approach has led to the proposal of legislation addressing various issues related to carbon market instruments. However, the concessions provided by the government to polluting industries significantly

---

[56] Beder above n 45.

[57] Lyster above n 53.

[58] Ibid.

[59] Stavins [16].

[60] Lyster, above n 53.

[61] Maraseni et al, above n 40.

[62] Guglyuvatyy [5].

[63] Ibid.

[64] Guglyuvatyy and Stoianoff above n 35.

influenced the design of the ACPRS to the point where the scheme's ability to reduce emissions were severely distorted.[65]

A notable success for the Rudd government was the introduction in 2009 of the Renewable Energy Target (RET) scheme. RET is expanding on the Howard government's mandatory renewable energy target. The RET legislation was enacted as Renewable Energy (Electricity) Amendment Act 2009. RET was designed to facilitate the government's commitment to ensure that 20% of Australia's electricity comes from renewable sources by 2020.[66]

Under the Renewable Energy Target Scheme, liable entities such as electricity retailers and large industrial users are required to purchase a certain percentage of their electricity annually from renewable sources.[67] As of January 1, 2011, the scheme has been split into two: the Large Scale Renewable Energy Target (LRET) for large renewable energy developments such as wind and solar farms, and the Small Renewable Energy Scheme (SRES) for small installations such as rooftop solar panels and solar water heaters.[68] Liable entities must fulfill their obligations by purchasing renewable energy certificates created by both large and small renewable energy producers. Typically, RET creates financial incentives for new or expanding renewable energy plants such as wind and solar farms.

The Clean Energy Finance Corporation (CEFC) and the Australian Renewable Energy Agency (ARENA), both established by the Rudd Government, have also been instrumental in assisting developers obtain funding by directly financing renewable energy projects and boosting private investment. These agencies have invested about $ 8.5 billion in clean energy projects, and these investments have spurred another $25–30 billion in additional private investment.[69]

3.3.1 Carbon Pricing

In 2010, there was a change of leadership and Kevin Rudd's deputy, Julia Gillard, became Australia's first female prime minister. Prime Minister Gillard announced the new government's intention to propose an interim carbon pricing scheme[70] and

---

[65] Crowley [3].

[66] Cludius Johanna et al. [8].

[67] Ibid.

[68] Ibid.

[69] ARENA (Australian Renewable Energy Agency) (2019), Annual Report 2018–19. Available at: https://arena.gov.au/assets/2019/10/arena-annual-report-2018-19.pdf   CEFC (Clean Energy Finance Corporation) (2019), Annual Report 2018–19. Available at: https://annualreport2019.cefc.com.au/

[70] A carbon pricing scheme is often called a 'tax' because during the fixed price period, the liable parties are obliged to purchase fixed price carbon units which is similar to paying tax. However, they cannot trade the units on the market, as under an emissions trading scheme.

create a multi-party climate change committee (the Committee)[71] made up of Commonwealth government members and senators. The intention of the Committee was to establish a climate change framework that would define the overall characteristics of the carbon price mechanism. The committee has published eleven policy principles to provide a consistent basis for carbon price discussions.[72]

The committee published the bill on July 28, 2011. In October 2011, the Australian House of Representatives passed the Carbon Pricing legislation, which was later passed by the Australian Senate. The Carbon Pricing Scheme, which was part of the Clean Energy Future regime, has been in effect since July 1, 2012 as a interim measure aimed at reducing greenhouse gas emissions.[73] The carbon price was $23 for the 2012–13 fiscal year and was expected to increase by 2.5% in each of the next two years.[74] Under this scheme, responsible businesses buy and surrender carbon units equal to their direct emissions (based on historical levels) of carbon dioxide (CO_2) equivalents. Failure to surrender the required carbon units incurs a penalty. Following the transition period, the carbon pricing mechanism was to be transformed into a cap and trade ETS providing flexible carbon pricing.[75] From July 1, 2015, carbon units were to be sold at auction. Consequently, even though the carbon pricing mechanism is sometimes referred to as a 'carbon tax', the Australian government has been committed to emissions trading.

The carbon pricing scheme covered four of the six greenhouse gases accounted for under the Kyoto Protocol, including carbon dioxide (CO_2), methane (CH_4), nitrous oxide (N_2O) and perfluorocarbon (PFCs), and broadly covered stationary energy, industrial processes sector, fugitive emissions (excluding decommissioned coal mines), and emissions from non-legacy waste.[76]

The scheme covered about 500 enterprises that emit 25,000 tonnes of CO_2 per year or more, and some waste plants that emit more than 10,000 tonnes per year, which is about 50% of Australia's greenhouse gas emissions.[77] Agricultural and transport fuels were excluded from the scheme. Fuel excise taxes have been increased for rail, domestic shipping and domestic aviation to reflect the price of carbon emissions. During the transition period with a fixed carbon price, liable enti-

---

[71] Guglyuvatyy above n 62.

[72] Ibid.

[73] Ibid.

[74] Ibid.

[75] Ibid.

[76] Stationary energy includes emissions from fuel consumption for electricity generation, fuels consumed in the manufacturing, construction and commercial sectors, and other sources like domestic heating. Industrial processes emissions are side-effects of production from non-energy sources, for example, it includes emissions from cement production, metal production, chemical production, and consumption of HFCs and SF6 gases. The fugitive emissions relates to the energy sector and covers emissions that are linked with the production, processing, transport, storage, transmission and distribution of fossil fuels such as black coal, oil and natural gas. The waste emissions relate to waste dumped at landfills.

[77] Zahar et al., above n 22.

ties were not allowed to use international emission reduction units. However, for the flexible pricing period, it was assumed that internationally recognised permits could be used to relieve up to 50% of a party's liability.[78]

During the period of fixed prices, there were no restrictions on emissions and the number of carbon credits was unlimited. It was expected that starting from 2015–16, the Climate Change Authority would set an emission cap in line with international and Australian emission reduction targets. The Gillard government has pledged to reduce emissions by 5% from 2000 levels by 2020 and 80% from 2000 levels by 2050.[79]

The carbon pricing scheme was supposed to generate about $24.5 billion in the first four years. However, the budget deficit was expected to be about $4 billion.[80] This is due to an extensive spending plan to support industries, households and investment in renewable energy. The government reformed the personal income tax and increased social security payments. The tax-free threshold for individuals was increased from $6000 to $18,200 from July 1, 2012. Consequently, all taxpayers with income below $80,000 have reduced their income tax payments.[81]

A $9.2 billion aid package was provided to Australian industries over the first three years to address competitiveness challenges posed by the carbon pricing scheme.[82] The hardest hit industries, such as steel, aluminum, zinc, pulp and paper, were expected to receive free permits covering about 94.5% of the industry's carbon costs. $300 million was to be set aside to facilitate the steel industry's transition to clean energy, and $1.3 billion was to be allocated to coal sector.[83] Additional measures were also considered to support research, development and commercialisation of green technologies. Clean Energy Finance Corporation was formed to invest $10 billion in new technologies, and $3.2 billion was assigned to the Australian Renewable Energy Agency.[84] The government has also pledged to shut down 2000 megawatts of dirtiest power generators by 2020.[85]

Since agricultural sector was not included in the carbon pricing mechanism, the government announced the Carbon Farming Initiative. The Carbon Farming Initiative is a carbon offset scheme that offers farmers, forest growers and

---

[78] Ibid.

[79] The Australian government has been criticized for these low targets to reduce greenhouse gas emissions. For example, Professor Garnaut (the Commonwealth government's climate change advisor) has recommended a 25% cut, while many other commentators suggest an even more ambitious target to reduce greenhouse gas emissions is needed. See for example: Garnaut, R. 2008. *Australia Counts Itself out*. Available: http://www.theage.com.au/national/australia-counts-itself-out-20081219-72ei.html?page=-1

[80] Guglyuvatyy above note 62.

[81] However, individual income tax rates for those with higher incomes were increased, for example: 19% for income over $18,200 (was 15%) and 32.5% for income over $37,001 (was 30%).

[82] Guglyuvatyy above n 62.

[83] Ibid.

[84] Ibid.

[85] Ibid.

landowners economic incentives to reduce carbon emissions.[86] Farmers and land-holders are able to create carbon credits via certain activities that could then be sold to other entities wishing to offset their carbon emissions. Particularly, the Carbon Farming Initiative has enabled participants to earn credits for actions such as: reforestation, reduction of methane emissions from the digestion of livestock, reduction of fertilizer contamination, protection of native forest, forest management, reducing pollution from legacy landfill waste and others.[87] Liable entities were also allowed to surrender Kyoto-compliant Carbon Farming Initiative credits up to 5% of their liabilities during the fixed price period, and there was no quantitative restrictions on usage of the Kyoto credits during the flexible pricing period.

The overall design of the carbon pricing scheme appeared to resemble in some respects the earlier proposed ACPRS.[88] In some respects, however, the carbon price was a significant improvement over the heavily compromised ACPRS. The significant compensation for the affected industries was temporary and based on historical emission levels, consequently incentives to reduce emissions were not undermined. The households assistance package was intended to compensate people with low and middle income. The increase in the income tax threshold has eliminated about a million low-income taxpayers from the income tax system. In addition, a number of supporting measures aimed at promoting carbon farming, energy efficiency and green innovation have significantly improved the ACPRS.

Overall, the Gillard government continued the Rudd government's efforts to mitigate carbon emissions but used a different tactic. Rather than enacting a direct emissions trading scheme, the government introduced the Clean Energy Future regime, which included an interim carbon pricing scheme, followed by an ETS. The Gillard government's multi-party climate change committee has refined the policy development process using clear principles that establish a consistent basis for carbon pricing. This approach was also a departure from the usual policy development process carried out by the relevant government departments.

Arguably, entrusting a Committee with the task of developing climate change policies is what has led to the successful implementation of Clean Energy Future regime. Carbon pricing characteristics offered significant improvements over the previously proposed ACPRS. The Clean Energy Future package also includes a number of new and effective measures enhancing carbon pricing mechanism. The Gillard government, with the support of the Greens and the Committee's policy-making approach, effectively overcame industry pressures. This led to the successful implementation of the Clean Energy Future by the Gillard government, although it was short-lived.

---

[86] Department of Agriculture Water and Environment. Carbon Farming Initiative at https://www.agriculture.gov.au/water/policy/carbon-farming-initiative

[87] Ibid.

[88] Guglyuvatyy above n 62.

3.4   Climate Change Mitigation 2013–2021

In September 2013, a coalition led by Tony Abbott won the Federal election. The new government's view on climate change was rather different from the views of the two previous prime ministers. The Gillard government's carbon pricing legislation was abolished by the new government and substituted with a Direct Action Plan by the end of 2014.

The Abbott government reviewed the Renewable Energy Target in 2013 to assess whether the goals were suitable and cost-effective.[89] The Renewable Energy Target set by the Rudd government has been lowered from 45,000 GWh to 33,000 GWh of electricity to be generated from renewable energy in 2015.[90] Regardless, Australia has met its 2020 renewable energy target of 23.5%, according to a report from the Clean Energy Regulator.[91] The share of renewables in the national electricity market was around 25% at the end of 2019.[92] Furthermore, a recent report from the Clean Energy Regulator indicates that renewables are likely to account for about 30% of all generation in the national electricity market by the end of 2021.[93]

The Abbott government declared the Direct Action Plan was designed to 'effectively and efficiently deliver low-cost emission reductions that will support our 2020 target.'[94] The Direct Action Plan includes an Emissions Reduction Fund (ERF) as a cornerstone to stimulate actions to reduce greenhouse gas emissions across the Australian economy. Funding from ERF is distributed through auctions, which means that the government pays for projects that will reduce CO_2 emissions at the lowest cost. The range of possible projects to reduce CO_2 emissions include: energy efficiency, cleaning up power plants, reforestation and vegetation restoration and others activities.[95]

Details were limited regarding the effectiveness and economic analysis of the Emission Reduction Fund and its structure.[96] There are many built-in issues with the Emission Reduction Fund. According to various commentators, the main issues that affect the ERF's ability to lower Australia's greenhouse gas emissions include: general emission limits, additionality, compliance mechanisms, penalties and access to international permits.[97]

---

[89] Alexander St John. The Renewable Energy Target: a quick guide. Parliament of Australia.

[90] Ibid.

[91] Clean Energy Regulator. The Renewable Energy Target 2019 Administrative Report – The acceleration in renewables delivered in 2019.

[92] Ibid.

[93] Clean Energy Regulator. Media release. 03 December 2021.

[94] The Emissions Reduction Fund. Available at: http://www.cleanenergyregulator.gov.au/ERF

[95] Ibid.

[96] The Australian Senate, Environment and Communications References Committee. 2014. Direct Action: Paying polluters to halt global warming?

[97] Ibid.

An important characteristic of the Direct Action Plan is its voluntary nature. The Voluntary Carbon reduction mechanism does not encourage businesses to participate and compete in the ERF.[98] The Australian Senate inquiry into the Direct Action Plan commented: 'The Committee is convinced that the Government Direct Action Plan and the proposed Emission Reduction Fund are fundamentally flawed. They ignore the well-established polluter pays principle and instead suggest that the Australian taxpayer should effectively subsidize the big polluters.'[99] Generally, the Direct Action Plan has been heavily criticised and described as a step backward in Australia's climate change policy.

One of the vital elements of the former Clean Energy Future package was the Carbon Farming Initiative. Carbon Farming Initiative has been retained under the ERF and has been enlarged to cover all sectors of the economy. Specifically, the ERF scheme has three elements: emission reduction credits, emission reduction purchases, and safeguarding emission reduction.[100] To obtain credits, the amount of the reduction provided by the project is first determined. Reductions must exceed business-as-usual activities. The regulator issues one Australian Carbon Credit Unit (ACCU)[101] for every tonne of emission reductions. ERF project participants can then sell their emission reductions (or ACCUs) to the Government through reverse auctions conducted by the Regulator.[102] The regulator buys the emission reductions at the lowest available price.[103]

Effective policies to reduce greenhouse gas emissions require limit or cap on total emissions and mechanisms to prevent sources of emissions from exceeding the limits. The direct action plan did not originally include emission limits or instruments to ensure that polluters limit their greenhouse gas emissions. To address this issue, a safeguard mechanism was announced in 2016. This mechanism applies to businesses that are required to report under National Greenhouse Gas and Energy Reporting regulations and emit more than 100,000 tonnes of CO_2-equivalent emissions in a fiscal year.[104] This applies to businesses across a wide range of industrial sectors, including power generation, mining, oil and gas production, manufacturing, transportation and waste. The safeguard mechanism requires the largest polluters in Australia to maintain their emissions within baseline levels. The safeguard mechanism covers 211 entities.[105] This mechanism ensures that the emission reductions

---

[98] See for example, Australian Government Department of Environment, Emissions Reduction Fund, public submission of Professor David Karoly, Professor Ross Garnaut, WWF-Australia and others. Available at: http://www.environment.gov.au/climate-change/emissions-reduction-fund/green-paper

[99] The Australian Senate above n 96.

[100] Carbon Farming Initiative Amendment Act 2014 (Cth).

[101] Australian Carbon Credit Units (ACCUs) are issued under section 162 of the Carbon Credits (Carbon Farming Initiative) Act 2011 (CFI Act 2011).

[102] The Emissions Reduction Fund above n 94.

[103] Ibid.

[104] National Greenhouse and Energy Reporting (Safeguard Mechanism) Rule 2015 (Cth).

[105] The Emissions Reduction Fund above n 94.

paid for through the ERF are not offset by significant increases in emissions from other sectors of the economy.[106]

The legislation is intended to confirm compliance with emissions limits by the covered entities. Emissions baselines provide a benchmark against which future emissions will be measured. The liable facility must maintain net emission levels at or below the baseline. According to the National Greenhouse and Energy Reporting Act 2007, all liable businesses must report their emissions to the Clean Energy Regulatory Authority by October 31 of each year.[107] The Clean Energy Regulator publishes information on all designated large entities emitting over 100,000 tonnes of CO_2 equivalent emissions each fiscal year (reference year).[108] Published information includes the baseline valid for that year, total reported emissions, responsible source (s) for each facility, and any carbon credits transferred in Australia.[109]

Baselines for existing entities are established using data provided under the National Greenhouse Effect and Energy Reporting System. The baseline reflects the highest reported emissions for the entity during the period 2009–10 to 2013–2014.[110] The safeguard mechanism also includes Flexible Compliance Arrangements that provide covered entities with access to a range of compliance options to meet their safeguard obligations. For example:

- A 'net emissions' approach allows businesses to use Australian Carbon Credit Units (ACCUs) to offset emissions.
- Long-term monitoring allows the businesses to exceed the baseline in one year, provided that the average emissions over two or three years are below the baseline.
- An exemption will be available for businesses whose emissions are a direct result of exceptional circumstances.
- There are a number of discretionary options that the Clean Energy Regulator can use to prevent non-compliance.[111]

These flexible compliance arrangements offer entities that are accountable under the safeguard mechanism a significant set of justifications for potential non-compliance with safeguard baseline emissions requirements.

The ERF is similar to the Kyoto Protocol's Clean Development Mechanism (CDM). The project must comply with the 'methodological definition' established by the Carbon Farming Initiative Law. The Minister of the Environment should consider:

---

[106] National Greenhouse and Energy Reporting Act 2007 (Cth); National Greenhouse and Energy Reporting (Safeguard Mechanism) Rule 2015 (Cth).

[107] Clean Energy Regulator. The safeguard mechanism. Available at: http://www.cleanenergyregulator.gov.au/NGER/The-safeguard-mechanism

[108] Ibid.

[109] Ibid.

[110] Ibid.

[111] Ibid.

- whether the definition meets the standards of offsets integrity;
- recommendations of the Emissions Reductions Assurance Committee; and
- whether any adverse environmental, economic or social impacts are likely as a result of the project.[112]

The 'offsets integrity standards' provisions entail that any methodology definition should result in a carbon dioxide emissions reduction that is improbable to occur in the normal course of business, including:

- removal of one or more greenhouse gases from the atmosphere;
- reduction of emissions of one or several greenhouse gases into the atmosphere;
- emission of one or more greenhouse gases into the atmosphere; and
- that removals, reductions or emissions can be measured and verified.[113]

Following the reporting period, a person may request the Regulator for the issuance of an entitlement certificate. The Regulator should ensure that the project adequately meets the 'additionality' requirements, including the requirement for novelty (the project has not started yet); the regulatory additionality requirement (the project is not implemented under another law); and the government program requirement (the project is not being implemented under another program/scheme).[114]

3.5   Emissions Reduction Fund Operation

The Emissions Reduction Fund was initially allocated a $2.55 billion budget for procurement of Australian carbon credits and is projected to deliver 240 MtCO$_2$e of emission reductions between 2021 and 2030.[115] In February 2019, the government provided an additional $2 billion in funding over the next 10 years, which is projected to lead to an additional reduction in emissions of 103 MtCO$_2$e by 2030.[116]

In April 2015, the Regulator conducted the first ERF auction. 107 carbon abatement contracts were awarded to deliver a total of 47,333,140 tonnes of greenhouse gas emissions reductions.[117] The total value of the contracts awarded was $660,471,500. The average price per tonne of abatement was $13.95. Carbon contracts have been awarded to 43 contractors for 144 projects.[118] The second ERF auction took place on November 4 and 5, 2015. 128 contracts were granted for the

---

[112] Carbon Credits (Carbon Farming Initiative) Act 2011 (Cth) Pt 10; ss 106, 123A.

[113] Carbon Credits (Carbon Farming Initiative) Act 2011 (Cth) s 133.

[114] Clean Energy Regulator. Emissions Reduction Fund. Available at: http://www.cleanenergyregulator.gov.au/ERF/Pages/default.aspx

[115] Ibid.

[116] Ibid.

[117] Clean Energy Regulator. Available at: http://www.cleanenergyregulator.gov.au/ERF/Auctions-results/april-2015

[118] Ibid.

purchase of 45,451,010 tonnes of abated emissions.[119] Contacts were granted to 77 contractors on 131 projects for a total amount of $556,875,549. The average price per tonne of abatement was $12.25.[120]

At the sixth ERF auction on December 7, 2017, the Regulator purchased 7.95 million tonnes of abatement. The average price per tonne of emission reductions was $13.08, which is the weighted average for each purchased ACCU. The overall value of all awarded contracts was $ 104 million, with the largest being 1.7 million tonnes and the smallest being 5000 tonnes.[121] The Environment Minister claimed that EFR was awarded a 180 million tonne contract to reduce emissions at less than $12 per tonne and it was quite 'cost-effective.'[122] The ninth auction was held in July 2019 at which the Clean Energy Regulator committed to purchase 59,000 tonnes of abatement at an average price $14.17 per tonne of abatement. Only three contracts were awarded for three projects, and the value of all contracts awarded was $840,000.[123] The 11th Emissions Reduction Fund auction was held in September 2020. The Clean Energy Regulator committed to purchase 7 million tonnes of abatement at average price $15.74 per tonne of CO_2 equivalent emissions.[124] In total, 35 contracts with total value of $110.3 million were awarded.

In 2021, the Clean Energy Regulator announced the registration of 145 ERF projects. This is a significant increase over previous years and the registration of soil carbon projects is the driving force behind the growth. Following COP2026,[125] when a number of countries, including Australia, committed themselves to be carbon neutral by 2050, Australia's carbon units' prices increased to $46 in December 2021 from $16 in early 2021.[126] While demand for carbon credit units is growing, the total supply of carbon units in Australia has also increased significantly to 17.3 million.[127] Australia's carbon market is hardly a 'free market' as the price of carbon units is highly dependent on the number of units issued by the Clean Energy Regulator. In 2021, with increased demand, the Clean Energy Regulatory approves more projects to increase the supply of carbon units and make it easier for big businesses to comply with the safeguard mechanism.

The results of the auctions demonstrated that the Emission Reduction Fund is not a sustainable policy as it does not provide permanent reductions in greenhouse gas

---

[119] Clean Energy Regulator. Available at: http://www.cleanenergyregulator.gov.au/ERF/Auctions-results/November-2015

[120] Ibid.

[121] Clean Energy Regulator. Available at: http://www.cleanenergyregulator.gov.au/ERF/Auctions-results/December-2017

[122] Hasham Nicole [12].

[123] Clean Energy Regulator. Available at: http://www.cleanenergyregulator.gov.au/ERF/Auctions-results/july-2019

[124] Clean Energy Regulator. Auction September 2020. Available at: http://www.cleanenergyregula-tor.gov.au/ERF/Pages/Auctions%20results/September%202020/Auction-September-2020.aspx

[125] UN Climate Change Conference. Available at: https://ukcop26.org/

[126] Clean Energy Regulator above n 93.

[127] Ibid.

emissions and a constant incentive for polluters to reduce their emissions. It is evident that the current government's (led by Scott Morrison after Tony Abbott's leadership as prime minister was challenged in October 2015) climate change policy is not adequate. The present government's policies have raised many questions and it is apparent that climate change policy in Australia has significantly deteriorated over the past years.

The current government's stance on climate change is fundamentally different from that of the two previous Labor governments. Following the Federal elections in September 2013, there was some skepticism about Tony Abbott's campaign pledge to end the carbon pricing scheme.[128] However, the carbon pricing legislation was abolished by the Abbott government shortly after the election. Some observers suggest that Tony Abbott 'reinstates industry influence over policy'.[129]

The government's Direct Action Plan to replace carbon pricing was heavily criticised from the outset. The Abbott government declared the Direct Action Plan was introduced to 'reduce emissions effectively and efficiently at low cost.'[130] The Abbott government did not disclose details regarding the development of the Direct Action Plan. It is not clear what the policy was based on. Successive Australian governments attempting to reduce greenhouse gas emissions have taken into account the developments and proposals of the previous governments. The policy implemented by the Abbott government is strikingly different from the carbon pricing and/or emissions trading mechanisms favored by the former Australian governments.

An important characteristic of the Direct Action Plan is its voluntary nature. The voluntary carbon mechanism does not encourage businesses to participate in the ERF. The Australian Senate inquiry into the Direct Action Plan made the following comment: 'The committee is convinced that the government's Direct Action Plan and the proposed Emission Reduction Fund are fundamentally flawed. They ignore the well-established principle of 'polluter pays', and instead propose that the Australian taxpayer should effectively subsidise big polluters.'[131]

The Emissions Reduction Fund is more of an artifice than an efficient measure to reduce emissions and encourage cross-sectoral participation. The main criticisms of the ERF are that it is not contributing to the Australian emission reduction target, it is not 'efficient' and 'ineffective.'[132] The scheme has been ridiculed as 'paying farmers to plant trees'.[133] A recent ERF review indicated that the scheme should cut less

---

[128] Crowley above n 65.

[129] Priest M. Coalition Eyes $20bn Carbon Cuts, Australian Financial Review, 9–10 March 2013.

[130] The Emissions Reduction Fund above n 94.

[131] The Australian Senate, Environment and Communications References Committee. 2014 'Direct Action: Paying polluters to halt global warming?

[132] Climate Change Authority, Review of the Emissions Reductions Fund (December 2017). Available at: http://climatechangeauthority.gov.au/review-emissions-reduction-fund

[133] Mark Ludlow, 'Carbon Farming Scheme Pays Money to Farmers to Plant Trees,' Australian Financial Review (6 May 2016). Available at: http://www.afr.com/news/politics/carbon-farming-scheme-pays-money-to-plant-trees-20160505-gona5e

emissions over time in Australia, and other policies will be needed to address the decarbonisation of the economy and bring about structural change.[134]

The Climate Change Authority's review found that the ERF is performing well overall but acknowledged that ERF alone is not enough for Australia to meet its emission reduction targets and other policies are required.[135] Clive Hamilton, a former member of the Board of the Climate Change Authority was scathing about the review, which lends a 'a veneer of legitimacy to a discredited scheme.'[136] The review did not report issues about additionality under the ERF.[137] Conversely, there is ample evidence that many such projects would have been implemented without ERF. For example, many landfill operators have joined the bidding for funds, even though they received finances from the Renewable Energy Target scheme.[138]

To date, most of the emission reductions under the Emission Reduction Fund are land based projects with nearly two-thirds of all emission reductions contracted to regrow vegetation or prevent land clearing.[139] One of the most significant problems with ERF is the substantial risk of reversal due to bushfires. For sequestration offset projects, a 5% 'risk of reversal buffer' and a 'permanence period' discount are applied to ensure the integrity of the ACCUs.[140] However, this is not enough. The bushfires in the summer of 2019 demonstrate that billions of dollars spent under the Emission Reduction Fund can be easily eliminated by bushfires in just a few months.

Economic theory argues that the use of economic instruments such as carbon taxes and ETSs outperforms subsidies. The disadvantage of the Direct Action Plan is its voluntary nature and inability to stimulate business participation in reducing emissions. ERF violates the principle that the polluter must pay. Australian taxpayers have effectively paid polluters for a scheme that is currently going up in smoke.[141]

In January 2019, the Organization for Economic Co-operation and Development (OECD) published a review of Australia's environmental performance, including the energy efficiency and carbon intensity of its economy. It emerged that Australia is one of the most carbon-intensive countries in the OECD, has adopted a fragmented approach to reducing emissions and must strengthen its climate change policies.[142]

---

[134] Hanna Emily. Climate change—reducing Australia's emissions. Commonwealth of Australia.

[135] Climate Change Authority above n 132.

[136] Hannam Peter, Bastard Child: Review of Controversial Emissions Funds Finds Significant Risk. The Sydney Morning Herald, 11 December 2017.

[137] Climate Change Authority above n 132.

[138] Hannam above n 136.

[139] Climate Change Authority above n 132.

[140] Carbon Credits (Carbon Farming Initiative) Act 2011 (Cth) Pt 2 Div 1.

[141] Hannam above n 136.

[142] OECD Environmental Performance Reviews: Australia 2019. OECD Environmental Performance Reviews, OECD Publishing, Paris.

3.6 Summary

Since the Howard government, the Commonwealth government has been addressing climate change and taking action. The efforts and attitudes of Australian governments varied significantly depending on the political parties that formed the government, and as a result, their policies varied significantly. The discernable explanation for such a different approach is the complex political games of the leading Australian political parties.

Australia's carbon policy relies heavily on reducing greenhouse gas emissions from land and forests but does not include forest protection as part of a climate change mitigation regime. The Emission Reduction Fund partially covers the forest and agricultural sectors. This policy may offer limited support to some landowners who regrow vegetation or prevent land clearing and maintain existing forests. Australia has various programs to protect forests and reduce emissions, but there is also a lack of coordinated efforts to tackle climate change. The next chapter examines some of the main current policies and issues related to the Australian governments' approach to climate change and forests.

References

1. Bamsey, H., & Rowley, K. (2015). *Australia and climate change negotiations: At the table, or on the menu?* Lowy Institute.
2. Beder, S. (2009). Token environmental policy continues in Australia. *Pacific Ecologist, 18*, 45–48.
3. Crowley, K. (2013). Irresistible force? Achieving carbon pricing in Australia. *Australian Journal of Politics and History, 59*(3), 368–381.
4. Gartrell, T. (2007). *Labor won the campaign outright, speech to the.* National Press Club.
5. Guglyuvatyy, E. (2012). Australia's carbon policy – A retreat from core principles? *eJournal of Tax Research, 10*(3), 552–572.
6. Guglyuvatyy, E., & Stoianoff, N. (2016). Carbon policy in Australia – A political history. In N. Stoianoff, L. Kreiser, B. Butcher, J. E. Milne, & H. Ashiabor (Eds.), *Green fiscal reform for a sustainable future – Reform, innovation and renewable energy.* Edward Elgar Publishing.
7. Howard, J. (2013). *One religion is enough, the global warming policy foundation annual lecture.* The Institution of Mechanical Engineers 5th November 2013. London.
8. Johanna, C., Forrest, S., & MacGill, I. (2013). *Distributional effects of the Australian Renewable Energy Target (RET) through wholesale and retail electricity price impacts.* Centre for Energy and Environmental Markets, The University of New South Wales.
9. MacGill, I., Passey, R., Nolles, K., & Outhred, H. (2005). *The NSW greenhouse gas abatement scheme: An assessment of the scheme's performance to date, scenarios of its possible performance to 2012, and their policy implications* (Discussion Paper DP050408). Centre for Energy and Environmental Markets.
10. McKibbin, W. J. (2007). Prime ministerial task group on emissions trading. *Agenda: A Journal of Policy Analysis and Reform, 14*(3), 13–17.
11. Narayan, M. T., Maroulis, J., & Cockfield, G. (2009). An analysis of Australia's carbon pollution reduction scheme. *International Journal of Environmental Studies, 66*(5), 591–603.
12. Nicole, H. (2017). Controversial Abbott-era climate fund will survive review: Josh Frydenberg. *The Sydney Morning Herald* (16 December 2017). Available at: http://www.smh.com.au/

federal-politics/political-news/controversial-abbottera-climate-fund-will-survive-climate-review-josh-frydenberg-20171214-h04pnq.html

13. Passey, R., MacGill, I., & Outhred, H. (2007). *The NSW greenhouse gas reduction scheme: An analysis of the NGAC registry for the 2003, 2004 and 2005 compliance periods* (Discussion paper DP070822). Centre for Energy and Environmental Markets.

14. Rosemary, L. (2008). The Australian carbon pollution reduction scheme: What role for supplementary emissions reduction regulatory measures? *University of New South Wales Law Journal, 31*(3), 880–893.

15. Staples, J. (2009). Australian government action in the 1980s. In H. Sykes (Ed.), *Climate change on for young and old*. Future Leaders, Melbourne.

16. Stavins, R. (2007). Proposal for a U.S. In *Cap-and-trade system to address global climate change: A sensible and practical approach to reduce greenhouse gas emissions*. The Brookings Institution.

17. Wilder, M., & Fitz-Gerald, L. (2009). Review of policy and regulatory emissions trading frameworks in Australia. *AERLJ, 27,* 1–22.

18. Zahar, A., Peel, J., & Godden, L. (2013). *Australian climate law in global context* (p. 156). Cambridge University Press.

Chapter 4
Australian Forests and Climate Change

Abstract This chapter examines the main forest laws and policies to assess the overall effectiveness of the existing regime. The efforts of the Commonwealth and the States to protect forests are considered with particular focus on Queensland. This analysis of the Australian regime allows us to trace the evolution of efforts to reduce greenhouse gas emissions, particularly in the light of the Paris Agreement signed by Australia. The analysis provides useful information but also allows to elaborate possible solutions and recommendations to improve Australia's climate regime, highlighting the need for coordination among all levels of government.

Keywords Deforestation · Federal and States' Forest policies and regulations · Australia's international commitments

4.1 Australian Forests

Australia's forests and woodlands are home to more than half of Australian terrestrial biodiversity and they act as major storage system for terrestrial carbon. Since European settlement in 1788, Australia has lost nearly 40% of its forests.[1] Australia's climate policy now relies heavily on forests and land use to achieve its mitigation goal.[2] Despite this dependence, approximately 50% of the remaining natural forests throughout Australia are estimated to be severely degraded.[3] Drought, fire, disease, pests and weeds are the main causes of deterioration in forest biodiversity and health, and stressors such as clearing, fragmentation and climate change are believed to be at the root of and exacerbate these problems.

Australia's forest laws have traditionally been administered by States and Territories. The Australian colonies were founded, developed responsible

---

[1] Bradshaw [5].

[2] Macintosh Andrew [3].

[3] Bradshaw above n 1.

governments and established formal forestry institutions and laws at different times. The first permanent settlement was New South Wales (1788) followed by: Tasmania (1803), Queensland (1824), Western Australia (1829), Victoria (1834), South Australia (1836). After political federation in 1901, these colonies became States of Australia, each with its own government, each retaining some of its former powers but surrendering others to the new Commonwealth (Federal) Government. Political structure of the country has significantly influenced the development of the forest laws in Australia. Amongst the powers the State Governments retained was control of the land within their borders and the forests on it. The significant part of the Australian legislation governing forests is State- or Territory-based, with the exception of a few Commonwealth Acts.[4]

The East coast of Australia, the State of Queensland in particular, is now ranked alongside countries such as Brazil, as a global front for deforestation.[5] In 2015 Australia was placed in the top 11 deforestation hotspots worldwide.[6] Deforestation and forest degradation considerably contributes to greenhouse gas emissions and it is practically impossible to stop climate change without reducing emissions from the forest sector.[7] The Australian climate change policy needs to be examined with reference to the forest policy to assess the most significant aspects of the current regime. Queensland's forest policy reviewed in further details to investigate high rates of deforestation in the State with largest forest area in Australia.

4.1.1 Deforestation in Australia

Deforestation is a direct contributor to climate change. Forests absorb carbon from the atmosphere, and when deforestation occurs, the carbon is released into the atmosphere. Globally, forests absorb about 7.6 billion tonnes of CO_2 annually, which is approximately 20% of global emissions.[8]

Australia has a total of 134 million hectares of forest, (around 17% of Australia's land area). Of this total forest area, 132 million hectares (98%) are native forests. Australia has about 3% of the world's forest area, and globally is the country with the seventh-largest forest area.[9] The largest area of Australia's forest is in Queensland (51.8 million hectares – 39% of Australia's forest), followed by the Northern

---

[4] Kanowski [17].

[5] World Wide Fund for Nature and International Institute for Applied Systems Analysis. 2015. Living Forests Report. Chapter 5: Saving Forests at Risk.

[6] Ibid.

[7] Huggins Anna [4].

[8] Harris et al. [13].

[9] Australian Bureau of Agricultural and Resource Economics and Sciences (ABARES) Australia's forests at a glance 2019. Canberra.

Territory (23.7 million hectares – 18%), Western Australia (21.0 million hectares – 16%), and New South Wales (20.4 million hectares – 15%).[10]

Although Australia has always had bushfires, the bushfire season in the summer of 2019–2020 was one of the most extensive in history. Bushfires affected about 8.5 million hectares of forests, including 8.3 million hectares of primary forests, 130,000 hectares of commercial plantations and 22,000 hectares of other forests.[11] However, human induced land clearing is the major factor contributing to deforestation in Australia.

Deforestation of low-reflectivity forests to replace them with higher-reflectivity crops or grassland[12] has negative impacts as it significantly increases atmospheric carbon levels worldwide.[13] This is because when trees are cut down and the root system is removed from the soil, carbon sequestration stops and the carbon stored in the trees and soil is released into the atmosphere as carbon dioxide.[14] The agricultural sector is dominated by industrial development, mainly driven by large farming enterprises. This often displaces smaller farms due to the power of large-scale enterprises. Agriculture is the major reason for clearing, with the greatest contribution from the activities of 'grazing native vegetation' and 'grazing modified pastures'.[15] Large-scale farms are 'continually convert[ing] wild lands into farms'[16] to increase income, which seriously affects the environment. The converted wild lands are commonly used to grow crops with higher reflectivity, grasses and livestock feed, or to increase livestock productivity.

Deforestation contributes to climate change by increasing greenhouse gas emissions. Forests are able to store significantly more carbon than agricultural fields, and forests are a critical component of the planet's carbon cycle. Soil is also a key factor in the planet's global carbon cycle, as it is the largest terrestrial organic carbon reservoir that acts as a source or sink for atmospheric carbon dioxide and is an important requirement for mitigation and adaptation to climate change.[17]

---

[10] Department of Agriculture, Water and the Environment. Australia's forests. Available at: https://www.agriculture.gov.au/abares/forestsaustralia/australias-forests

[11] Whittle [24].

[12] Ekström Marie et al. [20].

[13] Intergovernmental Panel on Climate Change. Sixth Assessment Report, 2021.

[14] Assefa Dessie et al. [12].

[15] Cullinan Cormac [7].

[16] Ibid.

[17] Soils can act as a major sink and source of atmospheric CO_2 and therefore have an important role in the carbon capture and storage activity.

4.2   Regional Forest Agreements

In the 1970s and 1980s, there was intense political debate about the future of the rainforest, particularly in New South Wales and Queensland.[18] Wood-chipping of native forests for export in 1970s sparked one of the most heated debates in Australian environmental history.[19] The debate focused on converting the ownership rights of commercial forests, managed by State agencies, into noncommercial tenure of National Parks.

In 1988, in response to this debate, National Forest Inventory (NFI) was created as an entity that enabled the calculation of nationally coherent attributes describing Australia's forests.[20] The NFI is directed by a steering committee, the National Forest Inventory Steering Committee (NFISC), which includes representatives from State, Territory and Australian government agencies involved in forest management.[21] The NFI management team at the Australian Bureau of Agricultural and Resource Economics and Sciences (ABARES) provides secretarial support to the NFISC.

The NFI aims to:

- be the source of integrated national forest data delivering multiple values,
- develop and endorse the use of national standards for the collection and management of forest data,
- provide information to support forest policy development, decision making, monitoring and reporting.[22]

The NFI accumulates on Australia's forests from State, Territory and Australian government agencies and integrates it into national classifications. The NFI is the key source of information about Australia's forests, including authorised national reporting requirements such as Australia' State of the Forests Report.[23] In 1992, the role of the NFI was recognised in the Australian National Forest Policy Statement (NFPS), which was signed by State and Territory governments and the Australian government.[24] The NFPS is intended to be initiated through the formation of a series of Regional Forest Agreements (RFA).

In the 1990s, debate over deforestation and logging of native forests intensified in several Australian States. Environmentalists pursued a ban on logging of native

---

[18] Davis [11].

[19] Ibid.

[20] McDonald Jan [15].

[21] Department of Agriculture Water and the Environment. Australia's National Forest Inventory. Available at: https://www.agriculture.gov.au/abares/forestsaustralia/australias-national-forest-inventory

[22] Ibid.

[23] Ibid.

[24] Ibid.

forests whereas the forestry industry pushed for greater security of resources.[25] In an attempt to resolve this issue the Commonwealth and State governments developed 20-year strategic plans. These plans, named Regional Forest Agreements (RFA), helped create a system of native forest reserves and delineate other areas earmarked for the forestry industry.

The RFA establishment allowed the Commonwealth government to delegate forest responsibility to the States and Territories. RFAs were intended to acknowledge diverse values in forest management and to balance economic and environmental considerations.[26] The major objectives of RFAs are to allow access to forest for the forest industry and to ensure sustainable management and conservation of Australia's native forests.[27] RFAs are a fundamental part of the management of the Australia's commercially productive native forests. The first RFAs were signed by States and Commonwealth governments between 1997 and 2001.[28] There are 10 RFAs covering commercial native forestry regions – five in Victoria, three in New South Wales, one in Western Australia and one in Tasmania.[29]

The introduction of RFAs has led to some notable advances in the management of native forests. Information on forest values has been significantly improved as a result of the comprehensive assessments. A large part of the forests was protected as National Parks instead being logged as woodchips.[30] About 70% of native forest in all regions were included in conservation reserves.[31] RFA and some other related initiatives resulted in a substantial increase in forest reserves. For example, the Western Australian government declared that in 2001 it will stop logging old-growth forests. Similarly, the 2005 Tasmanian Community Forest Agreement added 170,000 hectares of forests to conservation reserves.[32] This has led to significant improvements in the protection of biodiversity and cultural values. In addition, the State forest agencies and industry acquired resources for the planting of commercial forests.[33] Importantly, the RFAs development has also demonstrated that the Commonwealth government has a limited capacity to reduce forest reserves, while the States are able to increase the reserves.[34]

Many Regional Forest Agreements will soon be renewed or have recently been renewed. For instance, the New South Wales and the Commonwealth government have revised and extended the RFAs until 2039.[35] Unfortunately, the Commonwealth

---

[25] Judith Ajani [1].

[26] Lindenmayer David [10].

[27] Department of Agriculture Water and the Environment above n 21.

[28] Lindenmayer above n 26.

[29] Department of Agriculture Water and the Environment above n 21.

[30] Lunney Daniel and Chris Moon [9].

[31] Ibid.

[32] Macintosh Andrew [2].

[33] Lindenmayer above n 26.

[34] Ramsey A. Carr's green legacy is a black mark. The Sydney Morning Herald, July 2005.

[35] NSW Regional Forest Agreements. https://www.dpi.nsw.gov.au/forestry/regional-framework

government intends to renew RFAs without the necessary changes to include new information related to climate change and bushfires.[36] The information underlying regional forestry agreements dates back to the 1990s. At that stage, the problem of climate change was just beginning to emerge and the value of carbon storage in native forests as well as other relevant factors, were not taken into account in the development of the RFA.[37]

RFAs are said to have missed their targets as Australia's native forests are in much worse condition than they were many years ago when RFAs were first negotiated and signed.[38] Leading Australian experts argue that RFAs should not be renewed without new information on the value of native forests and on the threats to their preservation.[39] RFAs frameworks are based on incremental changes rather than catastrophic events such as bushfires. The RFAs do not include a mechanism for adjusting the terms of the agreement to change the balance between conservation and logging, which has become necessary due to bushfires.[40]

4.3 Deforestation in New South Wales and Victoria

About 900,000 hectares of State forest land were burned during the 2019–20 fires, according to a NSW government report.[41] Of the 522 State forests in New South Wales, 43 forests were almost completely burned down. A report released by the NSW Forestry Corporation confirms that fires have dramatically reduced the amount of timber that can be sustainably harvested.[42] Nonetheless, in Victoria and New South Wales, two of the worst-hit States by the fires, logging companies continue to cut native trees despite insensitive criticism from scientists and conservationists.

In 2017, New South Wales relaxed its native vegetation clearing laws. The NSW government implemented the Land Management and Biodiversity Conservation reform package, that comprised the new Biodiversity Conservation Act 2016 and amendments to the Local Land Services Act 2013. The reforms are being implemented in four key areas, namely:

---

[36] Lindenmayer David. Native forest protections are deeply flawed, yet may be in place for another 20 years. The Conversation, March 2018.

[37] Lindenmayer above n 26.

[38] Ibid.

[39] Lindenmayer above n 26.

[40] Blakers Margaret [19].

[41] NSW Government. NSW Fire and the Environment 2019–20 Summary. 2020 State of NSW and Department of Planning, Industry and Environment.

[42] 2019–20 Wildfires, NSW Coastal Hardwood Forests Sustainable Yield Review. https://www.dpi. nsw.gov.au/__data/assets/pdf_file/0004/1299388/fcnsw-sustainable-yield-report-2019-20-wild-fires.pdf

- The Land Management Framework (Natural Vegetation) Code, which defines the types of native vegetation clearing permitted on private land;
- $240 million investment in private land conservation managed by the Biodiversity Conservation Trust;
- improved frameworks for the management of native plants and animals, a process for protecting Areas of Outstanding Biodiversity Value and a modernised process for listing threatened plants and animals;
- the Biodiversity Offsets Scheme.[43]

The impact of this reform was evident in the reporting periods of 2019 and 2020. A Natural Resources Commission report indicates that land clearing has increased significantly and agriculture is the main driving force behind clearing in NSW. More than 37,000 hectares were approved to be cleared in 2018/19 in NSW (excluding clearing for invasive native species). This is about 13 times the pre-reform average annual approval rate, which was approximately 2700 hectares per year from 2006/07 to 2016/17.

Land clearing in NSW continued to grow after the NSW government relaxed vegetation laws, prompting expert and environmental groups to call for urgent revision of the legislation. Available data suggest that there is a serious risk of unexplained cleaning. Clearing of about 60% of agricultural land being unexplained. This, combined with a significant growth in the number of clearing approvals, poses a substantial risk to biodiversity and the legitimacy of reforms.[44] The government statistics shows 54,500 hectares of woody vegetation were cleared in 2019, marginally less than in 2018 and 2017, but 40% above the long-term average of 38,800.[45] Including non-woody vegetation, the clearing of regulated land in New South Wales reached 75,636 hectares, and 74% of these were deemed unexplained, that is, did not require an environmental permit or were illegal.[46]

In Victoria, more than 1.2 million hectares of land were burnt or affected by the fires, however the Regional Forest Agreement was lately renewed for 10 years, permitting the State's own logging company, VicForests, to manage logging in the State, including the critically endangered mountain ash forest ecosystem.[47] The Victoria government has declared that logging in native forests will stop by 2030, but environmentalists suggest that another 10 years of logging could result in ecosystem destruction. The North East Forest Alliance has challenged the State and Commonwealth governments in the Federal Court over the decision to roll over a

---

[43] NSW Government Natural Resources Commission. Review of land management and biodiversity conservation reforms. https://www.nrc.nsw.gov.au/land-mngt

[44] Ibid.

[45] NSW Department of Planning, Industry and Environment. 2019 landcover change reporting. https://www.environment.nsw.gov.au/topics/animals-and-plants/native-vegetation/landcover-monitoring-and-reporting/2019-landcover-change-reporting

[46] Ibid.

[47] Victorian Regional Forest Agreements. https://www.delwp.vic.gov.au/futureforests/what-were-doing/victorian-regional-forest-agreements

Regional Forest Agreement without taking necessary assessments of the environment.[48]

Australian experts established a direct link between the logging industry and wildfires, arguing that logging exacerbated the catastrophic wildfires that raged across Australia in the summer of 2019/20.[49] It is recognised that climate change has played a significant role in fuelling fires, but it is also argued that a more comprehensive discussion of land management and forestry practices is needed. Logging should be stopped in fire-prone areas and close to human settlements, and logging should focus on plantations rather than native forests. In March 2020, members of Wildlife of the Central Highlands (WOTCH), group of citizen scientists based in Victoria's Central Highlands, filed a lawsuit in the Supreme Court against VicForests for logging in unburnt habitat of rare and endangered species.[50] Represented by Environmental Justice Australia, WOTCH has had some success in this endeavour, with logging temporarily halted in 26 areas of the Central Highlands.[51]

4.4 Deforestation in Queensland

The largest area of Australian forests is located in Queensland.[52] Many more trees are lost in Queensland than are planted nationally by the Commonwealth government revegetation programs. Deforestation is a serious problem and could cause imminent damage to the Great Barrier Reef and the climate if not tightly regulated in the near future.[53] Agricultural land activities are considered to be the main driver of deforestation and land clearing in Australia,[54] with Queensland accounting for approximately 50–65% of total native forests loss over the past four decades.[55]

The tenant of the leased land has certain obligations to take care of the land.[56] When a lease is issued for agriculture, grazing or pastoral purposes, the lessee bears various responsibilities in relation to the duty of care for the land.[57] Given that the vast majority of land in Queensland is leasehold and used primarily for grazing

---

[48] North East Forest Alliance Challenges Regional Forest Agreements https://www.newsofthearea. com.au/north-east-forest-alliance-challenges-regional-forest-agreements-76452

[49] Lindenmayer et al. [18].

[50] WOTCH v VicForests (No 3) [2020] VSC 220.

[51] Ibid.

[52] Department of Agriculture, Water and the Environment above n 21.

[53] Cullinan Cormac [8].

[54] See, eg, Australian Bureau of Statistics, Agricultural Commodities, Australia, 2016/2016 (6 July 2017).

[55] Evans Megan C. Australia needs better policy to end the alarming increase in land clearing (9 August 2016) The Conversation. Available at: https://theconversation.com/australia-needs-better-policy-to-end-the-alarmingincrease-in-land-clearing-63507

[56] Land Act 1994 (QLD) s 199(1).

[57] Land Act 1994 (QLD) s 199(2).

pasture and cattle stations, it is on these properties where deforestation is most prolific.[58]

Illegal tree clearing is a serious issue in Queensland,[59] with some incidents receiving little attention.[60] Among the previous cases of illegal deforestation, a common excuse was that the clearing was done in error or that the permitted clearing areas were incorrectly defined.[61] Illegal clearing can lead to significant environmental impacts and it is highly questionable whether a person should be able to avoid responsibility for unlawful clearing by simply claiming that they have made a mistake, misunderstood their responsibilities,[62] or decided to clear the forest illegally, regardless of their right or permission to do this.

Deforestation restrictions were a critical requirement for Australia to meet its obligations under the Kyoto Protocol, however during the Newman Liberal National Party Government (Queensland) reign (2012–2015), some restrictions and laws were lifted and some repealed, leading to increased deforestation.[63] The resurgence of deforestation in Queensland is causing devastating effects on the Great Barrier Reef and water,[64] and is triggering significant national greenhouse gas emissions, putting additional pressure on Australia to meet its international commitments under Paris Agreement.[65]

---

[58] Willacy Mark. Strathmore Station land clearing under investigations for breaches of law, damage to environment (22 November 2015) Australian Broadcasting Company: ABC News. Available at: https://www.abc.net.au/news/2015-11-22/land-clearing-investigated-for-legal-breaches-environment-damage/6961108; McCutcheon Peter. Environment Department making 'urgent inquiries into clearing of Cape York land' Australian Broadcasting Company: ABC News. Available at: https://www.theguardian.com/australia-news/2017/jul/25/unapproved-land-clearing-an-unfolding-environmental-crisis-green-groups-say

[59] Robertson Joshua. Olivevale Station-Unapproved land clearing an unfolding environmental crisis, greens say (25 July 2017) The Guardian. Available at: https://www.theguardian.com/australia-news/2017/jul/25/unapproved-land-clearing-an-unfolding-environmental-crisis-green-groups-say

[60] Roberts George. Wombinoo Station owners fined for clearing 130 hectares of land without permission (19 December 2017) Australian Broadcasting Company: ABC News. Available at: http://www.abc.net.au/news/2017-12-19/wombinoo-station-owners-fined-for-land-clearing/9274154

[61] Ibid.

[62] Arthur Penelope and Lucy Kinbacher. North Queensland family seek action over ABC claims of illegal clearing (28 November 2017). Available at: https://www.illawarramercury.com.au/story/5088549/north-queensland-family-seek-action-over-abc-claims-of-illegal-clearing/

[63] Robertson above n 59.

[64] Ibid.

[65] Creary Marcia. 2013. Impacts of Climate Change on Coral Reefs and Marine Environment. L1 UN Chronicle 1–3. Available at: https://www.un.org/en/chronicle/article/impacts-climate-change-coral-reefs-and-marine-environment

4.4.1 Queensland Forest Regulation

The main piece of legislation governing land clearing and deforestation in Queensland is the Vegetation Management Act 1999 (Qld) (VMA). The initial implementation of the VMA resulted in a reduction in large-scale land clearing, which had been abundant in previous decades and which was identified as an important element in the early fight against land use impacts on the Great Barrier Reef.[66] Despite the Queensland Labor government's acknowledgment of significant environmental damage from the previous deforestation, the attempt to protect the land from clearing through the VMA was considered an ineffective approach by some commentators.[67] It was argued that the VMA is an obscure and irrational policy, introduced hastily.[68]

In 2013, during the Newman Liberal National Party Government reign, VMA that applied to restrict clearing of high value regrowth on privately held and indigenous lands were amended.[69] Moreover, the changes allowed broadscale clearing of 'high value' intensive agricultural production and self-assessment of areas that contain remnant or high value regrowth.[70] In 2015, with the change of State government, the Queensland Labor Party pledged to re-strengthen tree clearing laws because the regulation of tree clearing was recognised as an important step toward assisting in the longevity and health of the climate, the Great Barrier Reef, wildlife and human populations.[71] In particular, the Queensland Labor government proposed several restrictions to improve environmental protection through vegetation management laws, and in 2015 the recommended law was introduced to parliament.[72] Despite this attempt to reintroduce a system of responsible vegetation management, the submission was defeated in the Parliament in 2016.[73]

In anticipation of pending changes to deforestation laws by the Queensland Labor Government, a large majority of farmers began to panic clear, with large increases of clearing in Great Barrier Reef catchment areas equating to a rise from 100,000 hectares to 395,000 hectares.[74] In the 2015–16 clearing rate in Queensland was the highest since 2003–04 (490,000 hectares/year).[75] As a result, around 80% of

---

[66] Great Barrier Reef Marine Park Authority. Great Barrier Reef Outlook Report 2014, 164.

[67] Kehoe Jo [16].

[68] Will Steffen and Annika Dean [23].

[69] Vegetation Management Framework Amendment Act 2013.

[70] Ibid.

[71] Explanatory Memorandum, Vegetation Management and Other Legislation Amendment Bill 2018, 1.

[72] Ibid.

[73] Ibid.

[74] Department of Science, Information Technology and Innovation (Qld), Land cover change in Queensland 2015–16 Statewide Landcover and Trees Study Report SLATS study – Executive Summary 2016, 2.

[75] Ibid.

Australia's land use emissions in 2015 originated in Queensland.[76] This is equivalent to roughly half of the forest cleared in the Brazilian Amazon rainforest in 2016.[77] This increase caused a significant amount of sediment to reach the Great Barrier Reef, suffocating the living ecosystem and preventing photosynthesis.[78]

4.4.2 The 2018 Amendments to the Vegetation Management Act 1999 (Qld)

The policy objective of the Vegetation Management (Reinstatement) and Other Legislation Amendment Bill (Qld) was to amend the VMA and other legislation[79] to reinstate responsible laws regarding deforestation.[80] Upon re-election in 2017, the Queensland Labor Government reintroduced amendments to protect remnant and high conservation value non-remnant vegetation, amend development vegetation clearing codes, and align the definition of high value regrowth vegetation with the International definition of High Conservation Value.[81]

The amendments were introduced into Parliament on 8 March 2018 and the Bill was referred to the State Development, Natural Resources and Agricultural Industry Development Committee for detailed consideration.[82] The Bill was subsequently passed following the Committee inquiry and public consultation period, with amendments, on 3 May 2018.[83]

The recent amendments to the VMA are a positive step towards limiting deforestation in Queensland,[84] but the protections these amendments provide may not be very effective. VMAs appear to require further attention to reduce deforestation.[85] While other laws and regulations were amended at the same time,[86] the VMA is the piece of legislation that is the primary policy addressing land clearing.

---

[76] Ibid.

[77] Climate Council [6].

[78] Creary above n 65

[79] Water Act 2000 (Qld); Planning Regulation 2017; Planning Act 2016 (Qld).

[80] Explanatory Memorandum above n 71, 1.

[81] Ibid.

[82] Vegetation Management and Other Legislations Amendment Bill 2018.

[83] Ibid.

[84] Seelig Tim, Submission No 186 to State Development, Natural Resources and Agricultural Industry Development Committee, Vegetation Management and Other Legislation Amendment Bill 2018 (Qld), 21 March 2018, 1.

[85] Pointon Revel, Submission No 183 to State Department, Natural Resources and Agricultural Industry Development Committee, Vegetation Management and Other Legislation Amendment Bill 2018, 22 March 2018.

[86] Water Act 2000 (Qld); Planning Regulation 2017; Planning Act 2016 (Qld).

The amendments to the VMA have been an essential instrument to safeguard Queensland's remnant and high-value regrowth vegetation.[87] Tree thinning regulations are no longer viewed as a low-risk practices[88] and the amendments are directed at restoring a sustainable vegetation management system that aims to protect a total of 862,506 hectares[89] of 'high-value regrowth on freehold and indigenous land[s]'.[90] Approximately 405,000 hectares (47%)[91] of this land lies within the catchment area of the Great Barrier Reef.[92] Ancillary amendments have been made to the VMA, such as the exclusion of high value agriculture as the objective of a clearing application[93]; the self-assessed tree thinning codes removal[94]; and the expansion of the Reef riparian areas.[95]

Recent amendments to the VMA have been criticised[96] by farmers and politicians.[97] The vast majority of farmers in the agricultural sector openly contemplated that the amendments pose a threat to their land and livelihoods, with some arguing that 'if trees are left unmanaged on water courses [it may] … kill of grass which is the best thing for controlling erosion'.[98] Statewide protests were staged in Queensland in a bid to stop amendments which protesters said would 'devalue assets, prevent producers from expanding their businesses and lead to more fruit and vegetables being imported into Australia'.[99] While the agricultural sector appears to be unhappy with the amendments, there are clear signs that the amendments are a major step towards tackling Queensland's deforestation problems and, in turn, contribute to preserving the Great Barrier Reef and mitigating the effects of climate change.

The identification of potential loopholes in the VMA regarding vegetation management is a matter of concern. These loopholes will continue to allow significant

---

[87] Explanatory Speech, Vegetation Management and Other Legislation Amendment Bill QLD (2018).

[88] Ibid 417.

[89] Ibid 416.

[90] Ibid 415.

[91] Ibid 416.

[92] Ibid.

[93] Vegetation Management and Other Legislation Amendment Bill 2018 (Qld) cl 16.

[94] Explanatory Speech above n 87, 417.

[95] Vegetation Management and Other Legislation Amendment Bill 2018 (Qld).

[96] Trenton Akers and Sarah Vogler, Premier Annastacia Palaszczuk heckled in Rockhampton (10 May 2018) The Courier Mail. Available at: https://www.couriermail.com.au/news/queensland/queensland-government/premier-annastacia-palaszczuk-heckled-in-rockhampton/news-story/d369741d48c76f5073a3be003b630e67

[97] Sky News. Land Clearing laws could strangle Queensland crops and livestock: Farmers. SkyNews.com.au, 1 May 2018 (Brendan Smith: presenter).

[98] Burke Gail. Tightening of Queensland's land-clearing laws prompts farmers' fear for future generations. ABC News (Brisbane) 4 May 2018. Available at: http://www.abc.net.au/news/2018-05-04/land-clearing-laws-tightened-as-farmers-fear-for-future/9722416

[99] Ibid.

clearing of native vegetation[100] due to ambiguous language that allows new self-assessment codes to be generated from an incomplete list of matters that can be manipulated.[101] In addition, forage harvesting will continue to be seen as an adopted code for the clearing of vegetation[102] and has the potential to be a loophole for large-scale clearing. This is due to the fact that the amendments do not provide for a strict harvesting volume, and do not restrict forage harvesting only for officially declared periods of drought in accordance with the recommendations.[103]

An additional concern is the issue with mistake defense provision. The amendments were not intended to remove the application of the mistake of fact defense provision.[104] This defense allows any person who undertakes or mistakenly undertakes action such as illegal trees clearing on the basis of an honest and reasonable but mistaken belief that such act was acceptable, will not be criminally liable for the act or omission to any greater extent than if the real state of affairs was what the existing person was believed to be.[105] For example, if a person conducts or mistakenly conducts an act of tree clearing, when that person can demonstrate that his behavior was committed solely because of a mistaken belief, that person will not be criminally liable. However, this rule may be excluded by express or implied provisions included in the law relating to the subject matter.[106]

Although the Queensland Labor government is providing landowners with information that contains the necessary knowledge and resources regarding the new laws now envisaged by the VMA, the mistake of fact defense can still be viewed by landholders as a loophole. This is troubling. Perhaps if the VMA has removed this defense by incorporating explicit or implicit provisions into legislation, illegal tree clearing activities may be reduced due to the risk of prosecution.

This impending loophole could allow illegal clearing to continue and be carried out without the risk of criminal prosecution, especially if the landholder claims to have simply made a mistake. While the VMA has limited provisions that will 'ensure that occupiers and developers conduct robust due diligence prior to undertaking a clearing activity',[107] it remains debatable whether the liability of landowners, tenants and developers for illegal land clearing is indeed strong enough.[108]

Further concerns arise with the Property Map of Assessable Vegetation (PMAVs). PMAVs are certified maps showing the location and extent of areas with regulated and unregulated (exempt or non-exempt) vegetation. Prior to the amendments of the VMA, agricultural specialists and government representatives publicly

---

[100] Seelig above n 84.

[101] Vegetation Management (Reinstatement) and Other Legislation Amendment Bill (Qld) 2016 cl 4.

[102] Vegetation Management Act 1999 (Qld) s 19O(1)(a)(iv); s 22A(2)(f).

[103] Seelig, above n 84.

[104] Criminal Code 1899 (Qld) s 24.

[105] Criminal Code 1899 (Qld) s 24(1).

[106] Criminal Code 1899 (Qld) s 24(2).

[107] Vegetation Management (Reinstatement) and Other Legislation Amendment Bill (Qld) 2016, 10.

[108] Roberts above n 60.

recommended landowners to ensure they applied for and received PMAV before amendments were made that would impact future land development.[109] Landholders were encouraged to apply and advised that otherwise the landholder would risk losing any future development rights to the land. In addition, landholders were informed that the possession of PMAV ensures that high-value land will be subject to the amended Act.[110]

PMAVs have been authorised and released at the request of landowners. PMAV recognises land classified as category X as exempt from restrictions and subsequently excluded from the amendments. The amendments allowed for minimal changes to the classification of native vegetation, with unregulated vegetation being reclassified as regulated, although this is subject to an accepted development vegetation clearing code.

As discussed above, deforestation in Queensland is higher than in any other Australian State,[111] and under previous laws,[112] land cleared after 1989 (otherwise called Category X)[113] is permanently exempt from clearing regulation, regardless of the value of regrowth, or negative impact it generates for the biodiversity of the Great Barrier Reef.[114] To provide confidence to landowners within the existing regulatory framework, this exemption continues to apply regardless of the recent amendments. This is worrying because these areas are estimated to cover at least one-quarter of the area with advanced secondary re-growth, areas that are recommended to be protected rather than classified as exempt.[115]

Furthermore, another problem is that the actual extent of vegetated areas which are currently classified by a PMAV as exempt from regulation, are in fact remnant vegetation and/or high value regrowth, which cannot be clearly identified.[116] This information is a necessary requirement of the government in order to cautiously assess and determine the environmental impact that deforestation will have in the future. It is of serious concern that after the release of PMAV, it becomes difficult for the government to change the map. This is problematic because vegetation communities are changing, as is the scientific understanding of these areas, therefore it is imperative that PMAV can adapt and respond to these changes, but this was not reflected in the amendments.[117] Consequently, it is likely that deforestation is still

---

[109] Queensland Country Life. PMAVs deliver essential asset to protection says Marland. Queensland Country Life (Online) 14 October 2016. Available at: https://www.queenslandcountrylife.com.au/story/4229690/no-pmav-youre-a-bloody-idiot/

[110] Ibid.

[111] April E. Reside et al. [22].

[112] Vegetation Management Act 1999 (Qld).

[113] Vegetation Management Act 1999 (Qld) s 20AO.

[114] Vegetation Management Act 1999 (Qld) s 20AO; s 20CA(1–2).

[115] Mark Phelps. PMAVs under threat: Extreme greens flag new attack. Queensland Country Life (Online) 13 March 2018. Available at: https://www.queenslandcountrylife.com.au/story/5280656/thought-your-pmav-offered-certainty-think-again/

[116] Pointon above n 85, p3.

[117] Ibid.

possible due to PMAVs released before the amendment and could occur on both small and large properties.

A concern is the lack of recognition that PMAVs pose a potential risk and are likely to exacerbate unregulated deforestation. Ideally, the identification of this risk is required to reflect the change and therefore not recognise the X category as permanently tied to all PMVAs. Failure to protect significant forests and bushland, which are currently classified as land category X[118] is alarming and may still allow large-scale deforestation regardless of amendments. Furthermore, the current 50-meter buffer zone in the Great Barrier Reef catchment area may not be sufficient given that Category X lands are exempt from clearing restrictions, accordingly clearance can still be legally conducted within the catchment areas.[119]

Native vegetation, or terrestrial flora, plays a vital role in the health of Australia's environment. It is unique, varied and specially adapted to the climate of Australia.[120]

Deforestation threatens this terrestrial flora, especially those species that are limited to small remnants of native vegetation because they cannot survive in the altered habitat created by land clearing.[121] Native vegetation must be managed in an environmentally sustainable manner that contributes to 'its enduring ecological, economic, social, cultural and spiritual value.'[122] While this statement is mirrored the Commonwealth government vision for a framework on native vegetation,[123] the VMA does not reflect this vision.

The VMA does not fully support the prominence of the Great Barrier Reef and the protection of forests, rather it can be seen as undermining their value due to poorly structured legislation and governance systems. The VMA indirectly allows incremental clearing to continue indefinitely. A better approach would be if regulations were balanced with the incentives that provide economic opportunities for landowners. Since, land clearing in Australia, including Queensland, is mostly done on private land,[124] incentives should be provided to landowners to minimise land clearing. For example, an efficient carbon pricing mechanism can provide landowners with the opportunity to generate income by relinquishing valid land clearing rights in exchange for carbon offset payments.

---

[118] The Wilderness Society, Submission No 184 to Natural Resources and Agricultural Industry Development Committee, Vegetation Management and Other Legislation Amendment Bill 2018 (Qld), March 2018, 2.

[119] Ibid.

[120] COAG Standing Council on Environment and Water. 2012. Australia's Native Vegetation Framework. Australian Government – Department of Environment and Energy.

[121] Ibid.

[122] Ibid.

[123] Ibid.

[124] Evans above n 55.

4.4.3 *Vegetation Management Act 1999 (Qld) and Reef 2050 Plan*

The examples of shortages in the amendments provide an overview of areas where deforestation may continue. The amendments are viewed by many as a step forward towards reducing deforestation, but the VMA appears to require a complete overhaul when addressing issues that are directly related to the Great Barrier Reef and climate change. There are clear inconsistencies between the projections and recommendations of the Reef 2050 Plan[125] and the amendments to the VMA. The failure of the amendments to alter the way the VMA operates does not encourage limiting deforestation, and does not address current and emerging risks of climate change.

Despite scientific reports on the causes of climate change,[126] the rise of anthropogenic greenhouse gases continues to threaten the stability of the Earth's climate system and ecosystems such as coral reefs and, especially, the Great Barrier Reef.[127] The Great Barrier Reef is a World Heritage Area[128] It is the world's largest reef system and the biggest living structure on the planet, spanning 344,400 square kilometers,[129] off the east coast of Australia.[130] Deforestation poses a significant threat to the Great Barrier Reef due to gully erosion and climate change.[131] Gully erosion occurs when sediment from cleared forest areas is released into the reef catchment areas and then is washed directly into the Great Barrier Reef lagoons.[132] Trees play a critical role in the stabilising and absorbing greenhouse gases.[133] Thus, when trees are removed, more greenhouse gases are released into the atmosphere, rather than being absorbed by the trees or other natural sinks such as soil.[134] Gully erosion caused by deforestation in these catchment areas place great strain on the

---

[125] The Reef 2050 Plan. Available at: http://www.environment.gov.au/marine/gbr/long-term-sustainability-plan

[126] Nicholas H. Wolff, Peter J. Mumby, Michelle Devlin and Kenneth R. N. Anthony. 2018. Vulnerability of the Great Barrier Reef to climate change and local pressures. 24 GCB 1978, 1978–1979.

[127] Ibid.

[128] Environment Protection and Biodiversity Conservation Act 1999 (Cth) pt 3 div 1.

[129] Great Barrier Reef Marine Park Authority, Protecting the future of the Great Barrier Reef; Great Barrier Reef Region Strategic Assessment' Strategic Assessment. Available at: http://hdl.handle.net/11017/2866

[130] World Wide Fund for Nature and International Institute for Applied Systems Analysis, above n 339.

[131] Schaffelke, B., Collier, C., Kroon, F., Lough, J., McKenzie, L., Ronan, M., Uthicke, S., Brodie, J., 2017. Scientific Consensus Statement 2017. Scientific Consensus Statement 2017: A synthesis of the science of land-based water quality impacts on the Great Barrier Reef, Chapter 1: The condition of coastal and marine ecosystems of the Great Barrier Reef and their responses to water quality and disturbances. State of Queensland.

[132] Ibid.

[133] Assefa et al. above n 14.

[134] Ibid.

Great Barrier Reef due to increased sediment and artificial nutrients from farming being washed into gullys and then into the river and creeks systems that flow into inshore reefs.[135]

The future protection of the Great Barrier Reef has undergone extensive research[136] and assessment to identify threats that could assist immediate management prioritization of the Great Barrier Reef.[137] Management of the Great Barrier Reef is guided by plans, policies and legislation,[138] such as the Reef 2050 Plan which provides the policy framework for protecting and managing the Great Barrier Reef.[139] It sets clear actions, targets, objectives and outcomes designed to foster long-term sustainability of the reef.[140] The strategic goal of the Reef 2050 Plan, published in March 2015 by the Queensland and Commonwealth governments is to ensure that decisions regarding the Great Barrier Reef are consistent with its recommendations for sustainable reef management and ensure their long-term preservation for future generations. To this end, it includes specific guidelines for policymakers to apply when negotiating, developing or revising policies or other relevant programs and stipulates that all future actions that may affect the Great Barrier Reef should lead to an overall improvement and restoration of the health and resilience of the ecosystem.[141]

The weaknesses and exemptions of the VMA will continue to allow deforestation. In turn, this will allow continued sediment runoff and gully erosion of the Great Barrier Reef catchment areas. Agricultural practices located in the catchment areas will still be able to clear trees and land. Accordingly, there is no certainty that there will be an apparent reduction in nutrient, sediment and pesticide loads released into the lagoons and oceans of the Great Barrier Reef, regardless of future projections.[142] The Reef 2050 Plan recommends decision-makers to document how their decisions have contributed to the achievement of Reef 2050's goals and objectives to aid future reviews.[143] However, it appears likely that the VMA amendments would appease little, if any, of the Reef 2050 Plan recommendations. Whilst the fundamental outcomes of the Reef 2050 Plan are to overall protect the Great Barrier Reef, mitigate the effects of climate change, and limit sediment pollution and ecological pressures, these appear to go unnoticed in the amendments of the VMA.

---

[135] How does sediment affect the Great Barrier Reef? Available at: https://www.reefplan.qld.gov.au/resources/explainers/how-does-sediment-affect-the-gbr

[136] Great Barrier Reef Marine Park Authority, Great Barrier Reef Outlook Report 2014. Australian Government.

[137] Australian Government. Strategic assessment – Great Barrier Reef. Available at: https://www.environment.gov.au/protection/assessments/strategic/great-barrier-reef

[138] Great Barrier Reef Marine Park Act 1975 (Cth).

[139] The Reef 2050 Plan https://www.awe.gov.au/parks-heritage/great-barrier-reef/long-term-sustainability-plan

[140] Ibid.

[141] Ibid.

[142] Ibid.

[143] Ibid.

The VMA does not appear to provide direct protection to the Great Barrier Reef and does not meaningfully contribute to climate change mitigation. Despite concerns that statutory provisions such as environmental laws interfere with vested property rights by regulating land use, development and human activities,[144] these laws appear to be necessary to ensure the long-term sustainability of the Earth. It is imperative to limit human destruction, such as deforestation, to preserve the Great Barrier Reef. There are many possibilities and one of these may come from the Commonwealth government intervention over State made laws. Australian Courts also may be called upon more frequently to address environmental issues associated with deforestation and climate change.[145] In 2017, the State of Victoria enacted laws that now protect the Yarra River,[146] thereby recognising it as one living and integrated natural entity.[147] This recognition can provide hope for future acknowledgement of inherent rights to nature and protection of the environment in Australia.

4.5   Australian Governments' Forest and Climate Policies Dichotomy

There is a disharmony between the Commonwealth and State Governments concerning environmental laws.[148] The primary responsibility for protecting the environment generally rests with each individual State, while the Commonwealth's role is limited to regulating national environmental issues.[149] However, the High Court has previously ruled that the Commonwealth can legally enact legislation relying on the Commonwealth of Australia Constitution Act (the Constitution) if the matter is of international concern.[150]

The Commonwealth Government has additional environmental responsibilities as enshrined in the Environment Protection and Biodiversity Conservation Act 1999 (Cth). These responsibilities arise when there is a significant impact on an issue of national environmental importance that allows the Commonwealth to intervene.[151]

---

[144] See, eg, Great Barrier Reef Marine Park Act 1975 (Cth) s 38DD; Environment Protection and Biodiversity Conservation Act 1999 (Cth) pt. 3; Aboriginal and Torres Strait Islander Heritage Protection Act 1984 (Cth) div 1.

[145] Environmental Defenders Office. Landmark Challenge to Protect 20,000 ha of Native Vegetation. Available at: https://www.edo.org.au/2019/09/04/landmark-legal-challenge-saves-20000-hectares-of-native-vegetation/

[146] Yarra River Protection (Wilip-gin Birrarung murron) Act 2017 (Vic).

[147] Ibid s 1(a).

[148] Kanowski above n 4; Macintosh above n 32.

[149] National Environment Protection Council Act 1994 (Cth) sch Intergovernmental Agreement; Parliament of Australia. Commonwealth Environment Powers and Australian Federalism.

[150] Commonwealth v Tasmania (1983) 158 CLR 1.

[151] National Environment Protection Council Act 1994 (Cth) sch Intergovernmental Agreement.

These laws provide that a person should not take actions that[152]; has or will have a significant impact[153]; or is likely to have a significant impact[154]; on the world heritage values (natural heritage or cultural heritage[155]) of the declared World Heritage property.

The Great Barrier Reef is one of the listed World Heritage properties[156] with clear laws regarding approval of activities that will have a substantial impact on a declared World Heritage Property.[157] This provision raises concern about the State laws that are designed to purportedly limit deforestation, protect the Great Barrier Reef, limit gully erosion and climate change, but the effectiveness of these laws is questionable. This casts doubt on the reluctance of the Commonwealth Government to intervene, especially as the Great Barrier Reef is a World Heritage Property.

An effective framework is needed that protects the Great Barrier Reef and mitigates the effects of climate change. Though, this will be a major challenge if governance continues through the two tiers of governmental jurisdictions. For example, the VMA fails to directly afford protection to the Great Barrier Reef despite the formal recommendations of the Commonwealth Government in the Reef 2050 Plan which advises all legislators on the elements that need to be incorporated into legislation. Redefining the management of the Great Barrier Reef by reassessing management and conservation objectives and how they are to be achieved is imperative.[158]

Recognising the importance of the Great Barrier Reef and identifying the human factors underlying its degradation, including those factors that contribute to climate change, can provide an opportunity to develop proactive management measures supported by effective laws that are consistent with each other. Shifting from a passive ecosystem management plan to an active ecosystem intervention that enables the recognition and reduction of human drivers that are causing multiple pressures[159] on the Great Barrier Reef and climate change is essential. If this shift is not implemented, the actions and negative impacts caused by human activities such as deforestation, will lead to irreparable damage to the Great Barrier Reef and increase the pressure on the climate.

The Australian governments invest in programs to plant or protect native vegetation. These programs have emission reduction benefits as well as other environmental and agricultural benefits such as stabilising soils from erosion and providing shelters and windbreaks for livestock.[160] For example, the Commonwealth

---

[152] Environment Protection and Biodiversity Conservation Act 1999 (Cth) s 12(1).

[153] Ibid s 12(1)(a).

[154] Ibid s 12(1)(b).

[155] Ibid s 12(3).

[156] Ibid pt. 3 div 1 s 24(B)(C).

[157] Ibid s 12.

[158] Terry P. Hughes [14].

[159] Ibid.

[160] National Landcare Program. Available at: http://www.nrm.gov.au/national-landcare-program

government is investing $1 billion between 2018–19 and 2022–23 in Phase Two of the National Landcare Program.[161]

State and Territory governments are taking action to increase the number of trees, for example, the New South Wales Government has set a target to plant five million trees in Greater Sydney by 2030.[162] The Queensland Government established the $500 million Land Restoration Fund in 2018 and $8.4 million Aboriginal CarbonPlus Fund in 2017 to support carbon farming through land management activities, such as managing bush fires and reducing land clearing.[163] In October 2018, the Northern Territory Government released its Aboriginal Carbon Industry Strategy to support the establishment of emissions reduction projects on Aboriginal land.[164] The South Australian Government is committed to supporting carbon farming through a range of measures including innovative financing models for projects, linking carbon offsets with other environmental drivers and markets.[165] Despite the Australian Governments policies aimed to mitigate climate change the inconsistency between States and Commonwealth laws significantly undermines the Australian international commitments to reduce greenhouse gas emissions.[166]

The Commonwealth Government has spent millions of dollars to protect endangered species since 2014,[167] but these investments are effectively being wiped out by the loss of critical habitats through land clearing in Queensland and elsewhere. Many more trees have been lost in Queensland than have been planted nationwide by the Commonwealth Government's revegetation programs.[168] Thus, it can be argued that the Commonwealth government's carbon policy[169] is not credible as it does not include forest protection as part of Australia's climate change mitigation policy.

---

[161] Ibid.

[162] NSW Government. Available at: https://www.dpie.nsw.gov.au/__data/assets/pdf_file/0004/366052/five-million-trees-grant-program-guidelines.pdf

[163] Queensland Government. Available at: https://www.qld.gov.au/environment/climate/climate-change/land-restoration-fund/about

[164] Northern Territory Government. Available at: https://denr.nt.gov.au/__data/assets/pdf_file/0006/584439/Aboriginal-Carbon-Industry-Strategy_A4_Digital.pdf

[165] South Australia Government Available at: https://data.environment.sa.gov.au/Content/Publications/Carbon-Planting-in-SA.pdf

[166] Ajani J. 2008. Australia's Transition from Native Forests to Plantations: The Implications for Woodchips, Pulpmills, Tax Breaks and Climate Change. Agenda. A Journal of Policy Analysis and Reform, 15(3); Kanowski above n 4.

[167] Australian Government, Threatened Species Strategy Year One Report Australian Government. Available at: http://www.environment.gov.au/system/files/resources/dc0680d1-c280-4500-8cc3-b071fda69d34/files/threatened-species-strategy-year-one-report.pdf

[168] Landcare Australia, 20 Million Trees Landcare Australia. Available at: https://landcareaustralia.org.au/our-programme/20-million-trees/

[169] Carbon Credits (Carbon Farming Initiative) Act 2011; Carbon Credits (Carbon Farming Initiative) Regulations 2011. Clean Energy Regulator, Carbon Farming Initiative/Emissions Reduction Fund.

The permanent protection of forests will lead to immediate reductions in emissions from deforestation as well as degradation of natural carbon stores due to human activities.

The Carbon Farming Initiative and the Emissions Reduction Fund[170] only partially cover the landscape and agriculture sectors. This policy initiative could provide limited assistance to some landholders who protect and maintain existing native forests and their stored carbon. However, the Emission Reduction Fund eligibility requirements exclude a significant number of landholders from the scheme.[171]

Effective forest and land protection policies can be made possible if the Commonwealth Government relieves State and Territory governments of their responsibilities by adapting knowledge from all levels of governance into a single effective legal framework that takes precedence over State laws. The permanent protection of forests will directly benefit biodiversity and the conservation of natural carbon stores.[172] By protecting and restoring Australia's forests, we enable them not only to improve biodiversity and protect the Great Barrier Reef but also to make a considerable contribution to reducing GHG emissions in Australia.[173]

4.6   Australia's International Climate Commitments

Australian climate change policy relies heavily on land, forest and carbon farming to reduce greenhouse gas emissions. Australia' Kyoto commitment was 8% increase in GHG emissions between 2008 to 2012 based on 1990 emissions[174] and deforestation restrictions have been a critical requirement for Australia to meet its Kyoto target.

Deforestation related emissions were very high in 1990, especially in Queensland and New South Wales. Drought and declining commodity prices, as well as policy changes in most States in the early 1990s, led to a significant reduction in the rate of deforestation. As a result, by 1995, emissions from deforestation in Australia had decreased by 46%.[175] Thus, in the 1990th, Australia achieved significant reductions in emissions from deforestation without additional economic costs. Further

---

[170] Ibid.

[171] Clean Energy Regulator, Want to participate in the Emissions Reduction Fund? Available at: http://www.cleanenergyregulator.gov.au/ERF/Want-to-participate-in-the-Emissions-Reduction-Fund/Planning-a-project/Eligibility-and-newness

[172] Lindenmayer David. 2014. Forests, Forestry and forest management. in David Lindenmayer, Stephen Dovers, and Steve Morton (eds), Ten Commitments Revisited: Securing Australia's Future Environment. CSIRO Publishing.

[173] Brendan G. Mackey, Heather Keith, Sandra L. Berry, David B. Lindenmayer. 2008. Green Carbon: The Role of Natural Forests in Carbon Storage, A green carbon account of Australia's south-eastern Eucalypt forest, and policy implications. Australian National University.

[174] Parliament of Australia. Terms and Impacts of the Kyoto Protocol. Available at: https://www.aph.gov.au/About_Parliament/Parliamentary_Departments/Parliamentary_Library/Publications_Archive/CIB/CIB9798/98CIB10

[175] Macintosh above n 32.

measures to limit deforestation, especially land clearing regulations, were gradually introduced in Queensland and New South Wales between 1995 and 2009. These regulations were introduced mainly for land degradation purposes and reducing greenhouse gas emissions was not the main goal of these policies.[176] The impact of these initiatives is significant: the area of natural forest reserves doubled between 1990 and 2007.[177] Accordingly, Australia received another tranche of effortless offsets from forest management.

Australia ratified the Paris Agreement on November 6, 2016. Australia's emissions reductions targets were reset under the Paris Agreement. In its Nationally Determined Contribution, Australia set a target of 26–28% reduction in emissions from 2005 levels by 2030.[178] This corresponds to a range of acceptable emission levels of 435–447 Mt. CO_2-eq in 2030, including land use, land-use change, and forestry (LULUCF).[179] However, Australia's GHG in 2016/17 actually increased by 0.7% to 550.2 Mt. CO_2e.[180] There is considerable skepticism as to whether Australia would be on track to meet the Paris Agreement target, even taking into account reductions under the ERF.[181]

The Kyoto 'hot air'[182] credits for Russia and Ukraine discussed in the second chapter are similar to the offset received by Australia under Article 3.7 (2) of the Kyoto Protocol.[183] The Commonwealth government has indicated that it intends to use excess emission units from the Kyoto Protocol to meet the Paris Agreement target.[184] Many countries have criticised this transfer. However, the Australian government has included the transfer of Kyoto Protocol emission units in its latest

---

[176] For example, Native Vegetation Act 2003 (NSW); Vegetation Management Act 1999 (Qld); Vegetation Management and Other Legislation Amendment Act 2004 (Qld); Vegetation Management Bill 1999: Explanatory Notes; Vegetation Management and Other Legislation Amendment Bill 2004: Explanatory Notes; Australian Government and Queensland Government, Natural Heritage Trust Bilateral Agreement 1997; Australian Government and Queensland Government, Natural Heritage Trust Bilateral Agreement 2004; and Australian Government, National Strategy for the Conservation of Australia's Biological Diversity. 1996. Canberra: Commonwealth of Australia.

[177] Davidson John, Stuart Davey, Sharan Singh, Mark Parsons, Belinda Stokes and Adam Gerrand. 2008. The Changing Face of Australia's Forests. Australian Bureau of Rural Sciences, Canberra; Montreal Process Implementation Group for Australia, Australia's State of the Forests Report: Five-yearly report 2008.
Australian Bureau of Rural Sciences, Canberra.

[178] Australia's Nationally Determined Contribution. Available at: https://www4.unfccc.int/sites/ndcstaging/PublishedDocuments/Australia%20First/Australia%20NDC%20recommunication%20FINAL.PDF

[179] Ibid.

[180] Ibid.

[181] UN Environment. Emissions Gap Report 2018.

[182] For detailed discussion of the 'hot air' see: Alain Bernard, Sergey Paltsev, John M. Reilly, Marc Vielle and Laurent Viguier. 2003. Russia's Role in the Kyoto Protocol. MIT Joint Program on the Science and Policy of Global Change. Report 98.

[183] Kyoto Protocol Article 3.7 (2).

[184] OECD. OECD Environmental Performance Reviews: Australia 2019. Paris.

Climate Solutions Package.[185] Such a move would significantly reduce real emission reductions to 17–18% from 2005 levels by 2030.[186]

The National Greenhouse Gas Inventory shows that overall greenhouse gas emissions in Australia fell 1.4% in the year to March 2020.[187] Emissions outside the electricity sector were static, while emissions from the electricity sector fell 4%. Australia's current climate change mitigation policy has been described as 'highly insufficient'.[188] It is also noted that if all other countries were to follow Australia's current policy trajectory, warming could reach over 3 °C and up to 4 °C.[189]

Currently, the Australian government continues to rely on the Emissions Reduction Fund now re-named the 'Climate Solutions Fund', even though it does not provide any significant emissions reduction.[190] The ERF projects are dominated by land use sector abatements with a high risk of reversal through bushfires.[191] For example, the 2020 fire season has resulted in approximately 830 million tonnes of carbon dioxide equivalent emissions.[192] To date, most of the emissions reductions contracted under the Emissions Reduction Fund are from land-based projects, with almost two thirds of all emission reductions coming from vegetation restoration or land clearing prevention projects.[193]

New South Wales Energy and Environment Secretary Matthew Keane recently called for action to tackle climate change, recognising that the severity of the wildfires that engulfed the State in the summer of 2019/2020 was the result of high temperatures caused by climate change.[194] The Minister's call to arms in order 'to win the climate wars' by addressing energy indicated that 'taking action to reduce our emissions' is an economic opportunity.[195] Then on 22 December 2019, the Prime Minister of Australia, Hon. Scott Morrison, repeated, after strong criticism of

---

[185] Australian Government. 2019. Climate Solutions Package.

[186] Climate Analytics. 2019. Australian political party positions and the Paris Agreement: an overview.

[187] National Greenhouse Gas Inventory. Available at: https://www.industry.gov.au/data-and-publications/national-greenhouse-gas-inventory-march-2020

[188] Climate Action Tracker. Available at: https://climateactiontracker.org/countries/australia/

[189] Ibid.

[190] Climate Solutions Fund http://www.cleanenergyregulator.gov.au/csf/Pages/CSF-home.aspx

[191] Climate Action Tracker above n 188.

[192] The total net emissions of 830 Mt CO_2-e includes absolute emissions of around 940 Mt. CO_2-e, comprised of carbon dioxide emissions of 850 Mt CO_2-e, 81 Mt. CO_2-e of methane and 9 Mt. CO_2-e of nitrous oxide, as well as carbon dioxide sequestration equivalent to negative 110 Mt. CO_2-e resulting from recovery after this season's and previous seasons' fires. Estimating greenhouse gas emissions from bushfires in Australia's temperate forests: focus on 2019–20. https://www.industry.gov.au/data-and-publications/estimating-greenhouse-gas-emissions-from-bushfires-in-australias-temperate-forests-focus-on-2019-20#footnote-4-ref

[193] Climate Change Authority. Review of the Emissions Reduction Fund 2020.

[194] Hon. Matthew Kean. Smart Energy Summit Speech. 10 December 2019. Available at: https://mattkean.com.au/news/media/smart-energy-summit-speech

[195] Ibid.

his government's inaction on climate change, that the Commonwealth government's climate change policies will remain unchanged.[196]

In 2021, Scott Morrison announced that Australia commits to net zero by 2050 target, stating that technology, innovation and the private sector will drive Australia's efforts to reduce greenhouse gas emissions.[197] Furthermore, Australia and 141 other countries signed the Glasgow COP26 declaration to end deforestation by 2030.[198] The signatory countries include global deforestation hotspots such as Brazil, Indonesia and the Democratic Republic of the Congo, as well as countries with the largest forested areas including Russia, Canada, the United States and China. The Glasgow declaration is the latest in a series of analogous declarations. The Katowice COP24 pledge,[199] the New York Declaration on Forests[200] and Sustainable Development Goal 15[201] include comparable commitments to stop deforestation by 2030, but the rate of deforestation is still very high throughout the world.[202]

It is questionable whether Australia will be able to meet its emission reduction targets given the current disarray between government policies on climate and forests. This tension between governments in Australia demonstrates how fraught government policy responses have been while evidencing the long history of ever-changing climate change related initiatives and policies. There is ambiguity in the Constitution regarding the separation of power in environmental regulation, and Australian governments have attempted to clarify their roles and responsibilities regarding environmental issues.[203] Since the late 1960s, there has been a struggle between the Commonwealth and State and Territory governments to create mutually agreed institutional and policy mechanisms that would recognise the asserted domains of each level of government.[204]

In federations such as Australia, there is often some tension between different levels of government over the most suitable distribution of roles and responsibilities. Especially in an area of environmental policy where there is continuous change

---

[196] Dalzell Stephanie. Scott Morrison says he accepts criticism for Hawaii holiday during bushfires, apologises for any upset caused. Available at: https://www.abc.net.au/news/2019-12-22/prime-minister-scott-morrison-hawaii-holiday-bushfires/11821682

[197] Verrender I. Scott Morrison's climate change policy is being left behind by corporate action. Available at: https://www.abc.net.au/news/2021-04-26/business-climate-change-action-leaves-morrison-behind/100094616

[198] Glasgow leaders' declaration on forests and land use https://ukcop26.org/glasgow-leaders-declaration-on-forests-and-land-use/

[199] COP24: Key outcomes agreed at the UN climate talks in Katowice https://www.carbonbrief.org/cop24-key-outcomes-agreed-at-the-un-climate-talks-in-katowice

[200] Turning the New York Declaration on Forests to New York Action on Forests https://sdg.iisd.org/commentary/guest-articles/turning-the-new-york-declaration-on-forests-to-new-york-action-on-forests/

[201] Sustainable Development Goal 15 – Life on Land https://www.globalgoals.org/15-life-on-land

[202] The state of the world's forests. https://www.fao.org/state-of-forests/en/

[203] Peel and Godden [21].

[204] Department of Environment and Conservation, Report for Period December 1972 to June 1974. 1974 – Parliamentary Paper No 298.

as new scientific evidence emerges and public opinion demands that government and industry change the way they operate over the decades, these contradictions are becoming increasingly evident and difficult to resolve.

4.7 Conclusion

Australia's current climate change policy does not take into account the polluter pays principle and its effectiveness in reducing greenhouse gas emissions is problematic. Despite a number of climate change initiatives that have been used and sometimes rejected by the governments, the current Australian approach to reducing greenhouse gas emissions is inadequate to tackle climate change.

Deforestation is a global problem that exacerbates anthropogenic factors that are becoming stronger and more diverse. Australia's current government system is clearly fragmented and offers ineffective laws, aggravated by complexity, jurisdiction and legality issues, in most cases not addressing the impact of deforestation on climate change. Commonwealth tax regulations related to conservation are not a strong incentive to conserve land and forests. The tax system in Australia encourages income- producing and business-related activities, and the necessary qualities of a system that provide effective incentives for land conservation are not yet fully realised. The introduction of an appropriate stand-alone framework that provides consistent and explicit incentives for biodiversity conservation can significantly improve the existing system.

Forests governance and climate change regulations require immediate Commonwealth and State reform to ensure the longevity and health of the Great Barrier Reef and effectively address climate change issue. Efficient institutions, backed by a Commonwealth governance structure that offers robust forest protection and provides effective climate change mitigation and adaptation regulations, can help successfully reduce deforestation. Having uniform Federal law can be more beneficial in productively assessing scientific evidence, ensuring a coherent plan of action, managing compliance and influencing social norms at the national level, thereby reducing the potential for inconsistencies and conflicts between Commonwealth and States regulations.

The Commonwealth government should be empowered to enact environmental laws. The clear environmental power would enable the Commonwealth to set advanced national environmental standards. Explicit environmental power would allow the Commonwealth to take a strong leadership role in cases of poor environmental performance by States and Territories. The Commonwealth must be responsible for protecting the environment, not reserve the right to choose. If the Commonwealth is the only responsible government under the Constitution, a single set of environmental rules and powers can be established and enforced at the national level to help Australia effectively combat deforestation and climate change.

References

1. Ajani, J. (2007). *The forest wars*. Melbourne University Press.
2. Andrew, M. (2010). *Reducing emissions from deforestation and forest degradation in developing countries: A cautionary tale from Australia*. The Australia Institute.
3. Andrew, M. (2012). The Australia clause and REDD: A cautionary tale. *Climatic Change, 112*(2), 169–188.
4. Anna, H. (2007). Protecting world heritage sites from the adverse impacts of climate change: Obligations for states parties to the world heritage convention. *Australian International Law Journal, 14*, 121, 122.
5. Bradshaw, C. (2012). Little left to lose: Deforestation and forest degradation in Australia since European colonization. *Journal of Plant Ecology, 5*(1), 109–120.
6. Climate Council. (2018). *Land clearing and climate change: Risks and opportunities in the sunshine state*. Available at: https://www.climatecouncil.org.au/resources/qld-land-clearing-report/
7. Cormac, C. (2011). *Wild law: A manifesto for earth justice* (2nd ed.).
8. Cormac, C. (2017). Great barrier reef v Australian federal and state governments and others. In N. Rogers & M. Maloney (Eds.), *Law as if earth really mattered*. Routledge.
9. Daniel, L., & Moon, C. (2012). *An ecological history of Australia's forests and fauna (1770–2010)*. NSW Office of Environment and Heritage.
10. David, L. (2018). Flawed forest policy: Flawed regional forest agreements. *Australian Journal of Environmental Management, 25*(3), 258–266.
11. Davis, B. (1989). Wilderness conservation in Australia: Eight governments in search of a policy. *Natural Resources Journal., 29*(1), 103–113.
12. Dessie, A., Rewald, B., Sandén, H., Rosinger, C., Abiyu, A., Yitaferu, B., & Godbold, D. L. (2017). Deforestation and land use strongly effects soil organic carbon and nitrogen stock in Northwest Ethiopia. *Catana, 153*, 89.
13. Harris, N. L., Gibbs, D. A., Baccini, A., et al. (2021). Global maps of twenty-first century forest carbon fluxes. *Nature Climate Change., 11*, 234–240.
14. Hughes, T. P., Barnes, M. L., Bellwood, D. R., Cinner, J. E., Cumming, G. S., Jackson, J. B. C., Kleypas, J., van de Leemput, I. A., Lough, J. M., Morrison, T. H., Palumbi, S. R., van Nes, E. H., & Scheffer, M. (2017). Coral reefs in the anthropocene. *Nature, 546*, 87–88.
15. Jan, M. D. (1999). Regional forest (dis)agreements: The RFA process and sustainable forest management. *Bond Law Review, 11*(2), 295.
16. Jo, K. (2014). Environmental law making in Queensland: The vegetation management act 1999 (Qld). *Environmental and Planning Law Journal, 26*, 392, 393.
17. Kanowski, P. (2017). Australia's forests: Contested past, tenure-driven present, uncertain future. *Forest Policy and Economics., 77*, 56–68.
18. Lindenmayer, D. B., Kooyman, R. M., Taylor, C., et al. (2020). Recent Australian wildfires made worse by logging and associated forest management. *Nature Ecology and Evolution, 4*, 898–900.
19. Margaret, B. (2020). *No longer tenable bushfires and regional forest agreements*. Environmental Justice Australia.
20. Marie, E., Gerbing, C., Grose, M., Bhend, J., Webb, L., & Risbey, J. (Eds.). (2015). *Climate change in Australia technical report*. CSIRO, 23[3.1].
21. Peel, J., & Godden, L. (2005). Australian environmental management: A 'Dams' story. *UNSW Law Journal, 28*(3), 668–695.
22. Reside, A. E., Beher, J., Cosgrove, A. J., Evans, M. C., Seabrook, L., Silcock, J., Wenger, A. S., & Maron, M. (2017). Ecological consequences of land clearing and policy reform in Queensland. *Pacific Conservation Biology, 23*, 219.
23. Steffen, W., & Dean, A. (2018). *Land clearing and climate change: Risks and opportunities in the sunshine state*. Climate Council of Australia.
24. Whittle, L. (2020). *Analysis of effects of bushfires and COVID-19 on the forestry and wood processing sectors*. Australian Bureau of Agricultural and Resource Economics and Sciences.

Index

The manufacturer's authorised representative in the EU is Springer
Nature Customer Service Centre GmbH, Europaplatz 3, 69115 Heidelberg,
Germany. If you have any concerns regarding our products, please
contact ProductSafety@springernature.com

Printed and bound by CPI Group (UK) Ltd, Croydon, CR0 4YY

29/04/2026

02099458-0015